Ernst Schering Research Foundation Workshop 27
Gene Therapy

Springer-Verlag Berlin Heidelberg GmbH

Ernst Schering Research Foundation
Workshop 27

Gene Therapy

R.E. Sobol, K.J. Scanlon, E. Nestaas
Editors

With 14 Figures and 20 Tables

Series Editors: G. Stock and U.-F. Habenicht

ISSN 0947-6075

CIP data applied for

Die Deutsche Bibliothek – CIP-Einheitsaufnahme
Schering-Forschungsgesellschaft <Berlin>: Ernst Schering Research Foundation Workshop. - Berlin; Heidelberg; New York; Barcelona; Budapest; Hong Kong; London; Milan; Paris; Santa Clara; Singapore; Tokyo: Springer.
ISSN 0947-6075
27. Gene Therapy. - 1998
Gene therapy: with 20 tables / R.E. Sobol ... ed. - Berlin; Heidelberg; New York; Barcelona; Hong Kong; London; Milan; Paris; Singapore; Tokyo: Springer, 1998
(Ernst Schering Research Foundation Workshop; 27)
ISBN 978-3-662-03579-5 ISBN 978-3-662-03577-1 (eBook)
DOI 10.1007/978-3-662-03577-1

Originally published by Springer-Verlag Berlin Heidelberg New York in 1998.
Softcover reprint of the hardcover 1st edition 1998

Typesetting: Data conversion by Springer-Verlag

SPIN: 10691219 13/3135–5 4 3 2 1 0 – Printed on acid-free paper

Preface

This book is a collection of preclinical and clinical reports on the application of gene therapy to human disease. The focus of these studies is on cancer and cardiovascular disease.

There are two fundamental technologies for delivering therapeutic genes to diseased cells: either viral vectors, as discussed by Dr. Balmain, or non-viral vector systems, as discussed by Dr. Felgner. The strengths and limitations of each of these delivery systems are characterized. The use of a therapeutic gene to treat a disease has taken two general approaches. The first is to introduce a normal (i.e., wild type) gene into the patient that will restore normal gene function. Dr. Weissman has characterized the tumor suppressor gene (p53), and has shown that it can restore normal cell function in cancer cells. The second approach is to treat the disease with antisense molecules. Abnormal gene expression can be down-regulated and selectively inhibited by antisense molecules, which can reverse the pathologic process in cancer cells. Dr. Gewirtz has demonstrated this with antisense genes on leukemia, while Dr. Scanlon has applied this principle using ribozymes in human carcinomas.

During this symposium, Dr. Engler described clinical studies of gene therapy using growth factors to stimulate new blood vessels in patients with cardiovascular disease. Several gene therapy strategies were used for cancer: overcoming drug resistance by Dr. Bertino, a pro-drug strategy with ganciclovir by Dr. Scardino, *ex vivo* gene therapy for hematopoietic cells by Prof. Mertelsmann, and immuno-gene therapy by Dr. Sobol. Finally, Dr. Wikberg-Leonardi discussed the regulatory issues governing gene therapy in Europe and the United States.

The purpose of this volume is to describe the current applications of gene therapy to the clinical treatment of human diseases, and to demonstrate the rapid maturation of this field. It is hoped that this symposium will further assist the understanding and development of a new category of therapeutic agents.

E. Nestaas
K. J. Scanlon

Table of Contents

List of Editors and Contributors

Editors

R.E. Sobol
Sidney Kimmel Cancer Center, 10835 Altman Row, San Diego, CA 92121, USA

K.J. Scanlon
Cancer Research Department, Berlex Biosciences, 15049 San Pablo Avenue, Richmond, CA 94806, USA

E. Nestaas
cmc/BRD, Berlex Biosciences, 15049 San Pablo Avenue, Richmond, CA 94806, USA

Contributors

H.L. Adler
Matsunaga-Conte Prostate Cancer Research Center and the Scott Department of Urology, Baylor College of Medicine, 6560 Fannin Street, STe. 2100, Houston TX 77030, USA

B. Anderegg
Cancer Research Department, Berlex Biosciences, 15049 San Pablo Avenue, Richmond, CA 94806, USA

A. Balmain
Onyx Pharmaceuticals, 3031 Research Drive, Richmond, CA 94806, USA

D. Banerjee
Program of Molecular Pharmacology and Experiment Therapeutics, Memorial Sloan-Kettering Cancer Center, 1275 York Avenue, New York, NY 10021, USA

R. Bartholomew
The Immune Response Corporation and Sidney Kimmel Cancer Center, Clemma A. Hewitt Gene Therapy Laboratory, 10835 Altman Row, San Diego, CA 92121, USA

J.R. Bertino
Program of Molecular Pharmacology and Experiment Therapeutics, and Department of Medicine, Box 78, Memorial Sloan-Kettering Cancer Center, 1275 York Avenue, New York, NY 10021, USA

R. Engler
Collateral Therapeutics, 9360 Towne Center Drive, San Diego, CA 92121, USA

H. Fakhrai
Sidney Kimmel Cancer Center, Clemma A. Hewitt Gene Therapy Laboratory, 10835 Altman Row, San Diego, CA 92121, USA

P.L. Felgner
Gene Therapy Systems, Inc., 3525 John Hopkins Court, San Diego, CA 92121, USA

M. Garrett
Sidney Kimmel Cancer Center, Clemma A. Hewitt Gene Therapy Laboratory, 10835 Altman Row, San Diego, CA 92121, USA

A.M. Gewirtz
Departments of Medicine and Pathology, Institute for Human Gene Therapy, and the Cancer Center Stem Cell Biology/Transplantation Program, Stellar-Chance Labs, University of Pennsylvania School of Medicine, 422 Curie Boulevard, Philadelphia, PA 19104, USA

H. Glimm
Terry Fox Laboratory, 601 West 10th Avenue, Vancouver, BC, V5Z 1L3, Canada

D.P. Gold
Sidney Kimmel Cancer Center, Clemma A. Hewitt Gene Therapy Laboratory, 10835 Altman Row, San Diego, CA 92121, USA

M. M. Gottesman
Intramural Research, Building 1, Room 114, National Institutes of Health, Center Drive, MSC 0140, Bethesda, MD 20892-0140, USA

C. Heise
Onyx Pharmaceuticals, 3031 Research Drive, Richmond, CA 94806, USA

R. Henschler
Department of Internal Medicine I, Albert-Ludwigs-University, Hugstetterstrasse 55, 79106 Freiburg, Germany

A. Irie
Cancer Research Department, Berlex Biosciences, 15049 San Pablo Avenue, Richmond, CA 94806, USA

D.H.Kirn
Onyx Pharmaceuticals, 3031 Research Drive, Richmond, CA 94806, USA

R. Mertelsmann
Department of Internal Medicine I, Albert-Ludwigs-University, Hugstetterstrasse 55, 79106 Freiburg, Germany

I. Royston
Sidney Kimmel Cancer Center, Clemma A. Hewitt Gene Therapy Laboratory, 10835 Altman Row, San Diego, CA 92121, USA

K.J. Scanlon
Cancer Research Department, Berlex Biosciences, 15049 San Pablo Avenue, Richmond, CA 94806, USA

P.T. Scardino
Scott Department of Urology, Baylor College of Medicine, 6560 Fannin Street, STe. 2100, Houston, TX 77030, USA

D. Shawler
Sidney Kimmel Cancer Center, Clemma A. Hewitt Gene Therapy Laboratory, 10835 Altman Row, San Diego, CA 92121, USA

R.E. Sobol
Sidney Kimmel Cancer Center, 10835 Altman Row, San Diego, CA 92121, USA

N. Takebe
Program of Molecular Pharmacology and Experiment Therapeutics, and Department of Medicine, Memorial Sloan-Kettering Cancer Center, 1275 York Avenue, New York, NY 10021, USA

C. Van Beveren
Sidney Kimmel Cancer Center, Clemma A. Hewitt Gene Therapy Laboratory, 10835 Altman Row, San Diego, CA 92121, USA

C. von Kalle
Department of Internal Medicine I, Albert-Ludwigs-University, Hugstetterstrasse 55, 79106 Freiburg, Germany

B.E. Weissman
University of North Carolina-Lineberger Cancer Center, University of North Carolina, 102 Mason Farm Road, Chapel Hill, NC 27599-7295

R. Wikberg-Leonardi
Collateral Therapeutics, 9360 Towne Drive, San Diego, CA 92121, USA

S.-C. Zhao
Program of Molecular Pharmacology and Experiment Therapeutics, Memorial Sloan-Kettering Cancer Center, 1275 York Avenue, New York, NY 10021, USA

1 Human Gene Therapy: Future Prospects

M. M. Gottesman

"Ah, but a man's reach should exceed his grasp, or what's a heaven for!"
Robert Browning, 1855

1.1 Introduction: Expectations and Reality

Human gene therapy has been both a disappointment and a triumph. It has disappointed in the sense that the early promise of safe and efficient treatment and cure of human disease has not come to pass several years after the first introduction of DNA into patients for therapeutic purposes. It has been a triumph because it has captured the creative energies and imagination of thousands of scientists throughout the world, has garnered extensive investment by governments, non-profit research foundations, and the private sector, and has provided a rational outlet for translation of discoveries in basic biology to the prevention, treatment, and cure of human disease.

This discussion will be limited to somatic gene therapy, defined as the introduction of DNA by a variety of means into cells of an organism, not including germ cells. The purpose of such treatment is not to change the overall genotype, but to alter the properties of individual cells or groups of cells in a way which improves the ability of the patient to deal with a disease, either genetic or acquired, thereby improving survival and quality of life. Such treatment is consistent with the goals of medicine, and can be embraced with enthusiasm by most people. Human gene therapy for cosmetic purposes, or to alter normal, as opposed to disease, phenotypes (such as to enhance intelligence, beauty, or control behavior), is not within the purview of this essay.

In the few years that human gene therapy has been defined and practiced we have come to realize that the unrealistic expectation of being able to visit the doctor's office, receive an injection of a vector which benignly targets specific DNA sequences to specific cells or tissues, and walk out of the office forever cured of a generalized disease is not currently within our grasp (Anderson 1995). This dream has been replaced by the more appropriate view of the complex technology needed for delivery of DNA sequences, far more limited expectations of short-term success in localized disease, and a much better concept of our areas of ignorance and the need for more basic research. Some of this "coming of age" of human gene therapy was crystallized in a *Report and Recommendations of the Panel to Assess the NIH Investment in Research on Gene Therapy* (Orkin and Motulsky 1995). This does not mean that our long-term vision for safe, efficient, and effective human gene therapy has altered, but only that our expectations for the near term have changed dramatically. In the discussion that follows, I will try to distinguish near-term prospects from longer term goals.

1.2 Areas of Opportunity in Gene Therapy

It has become apparent that some of the most serious problems in the field of human gene therapy are really opportunities for new discoveries and approaches in biomedical research. Several such opportunities which have already resulted in success or may soon do so are highlighted below:

1. Human gene therapy is itself a tool for the study of human disease. Some of the earliest success in gene therapy came from so-called gene marking studies to demonstrate the feasibility of introducing DNA sequences into human cells in patients. The best designed of these early clinical trials answered important biological questions which would have been difficult to answer without the use of gene marking techniques. For example, Brenner and colleagues, by genetically marking leukemia cells in pediatric patients, showed that recurrence of leukemia in patients receiving autologous bone marrow transplants was from the transplanted tissue, rather than residual disease in the patient's surviving bone marrow (Brenner 1996). Such studies focused efforts to improve purging of bone marrow used in autologous transplants.
2. The need for preclinical animal models to test human gene therapy has spurred development of mouse monogenic insertional mutants. Although the power of mouse genetic models of human disease has long been recognized, the desire to have systems in which gene therapy technology could be tested has resulted in some models which have expanded understanding of disease pathophysiology. For example, a glucocerebrosidase knockout mouse, developed as a model for Gaucher disease, was found to have a previously unrecognized, early lethal phenotype including ichthyosiform skin, which was subsequently confirmed to be characteristic of complete loss of the enzyme in humans (Sidransky et al. 1996).
3. Human gene therapy has stimulated clinical research activity. The excitement generated by the prospect of human gene therapy has piqued the interest of early career investigators and attracted them to a field in which basic research could be translated readily into patient-oriented research. Both clinicians and basic laboratory researchers have become interested in human gene therapy clinical trials. There have been over two hundred such trials in the U.S. alone, with many also underway in Europe and Japan. At a time when academic medical centers are having increasing difficulty finding the resources to undertake innovative clinical trials, human gene therapy has attracted investment by pharmaceutical firms and

biotechnology companies and some of the best minds in biomedical research.

4. Human gene therapy has forged new partnerships between government, academia, and private industry. Although the alliance of private investors, whose bottom line may be profits, and academics, who are interested in enhancing knowledge and improving medical care, may occasionally generate some conflicts, there can be no question that the involvement of these three sectors has opened up opportunities for new research which would otherwise not have been available.
5. Research on gene therapy has stimulated basic biomedical research and the development of new technologies. Numerous examples exist of improvements in understanding of vector systems, the basis of viral specificity and pathogenesis, mechanisms of recombination and maintenance of extrachromosomal DNA, development of cationic and other liposome systems for DNA delivery, mechanisms of action of anti-sense DNAs and ribozymes, etc. Basic laboratory science in these areas has been accelerated by the desire to use these technologies for more efficient human gene therapy.
6. Early gene therapy experiments may prove concepts about the pathogenesis or treatment of human disease which can be exploited to improve therapy by means other than gene therapy. For example, if some strategies to treat cancer, which seem unlikely to be efficient enough to work as gene therapy, such as introduction of a tumor suppressor, a ribozyme, anti-sense oligonucleotides, or a dominant negative protein fragment to telomerase or to an oncogene into a specific malignancy result in regression of this tumor, then this pathway would be an appropriate target for drug development. Thus, although in many cases gene therapy per se may be impractical, information gleaned from gene therapy clinical trials can be useful in the development of other therapeutics.

1.3 Short-Term Goals

The clinical research which has already been done on gene therapy has helped to clarify what experiments are likely to succeed and what experiments are unlikely to be fruitful. In this section, several examples will be provided of goals which seem feasible in the next few years, based on current experience.

One lesson of the early gene therapy experiments was the need for more careful experimental design. Uncontrolled experiments in gene therapy, like uncontrolled experiments in any branch of experimental science, are doomed to failure, or at least to endless argument about which variable was responsible for the effect observed. Thus, the groundbreaking experiments of Blaese, Rosenberg, Anderson and their colleagues, in which the introduction of an adenosine deaminase (ADA) cDNA into the peripheral lymphocytes of two children with ADA deficiency resulted in apparent expression of this cDNA in one child and improved immunologic status in the other (Blaese et al. 1995), cannot unequivocally be proven to be the result of the experiment, since PEG-ADA given to these children has also been shown to be a potent immunologic restorant, and the procedure of in vitro amplification of lymphocyte populations in itself might be salutory. Given the limited number of patients with ADA deficiency, and the pioneering nature of these studies, it is unlikely that the experimental design could ever have been adequate to answer all critics, but this case points up the need for better controlled experiments.

The second point made by the ADA studies is the need for careful evaluation of the subjects to get the most possible information from our clinical trials. Blaese and coworkers painstakingly followed the presence and expression of the ADA cDNA in both of their patients for several years, looked for development of antibodies to neomycin phosphotransferase, the selectable marker used in vitro to select for transduced lymphocyte populations, and carefully evaluated all aspects of immunologic status of their patients. Thus, although the trial design was not perfect, much useful information was gleaned from these studies. This principle of the importance of careful Phase I analysis of gene therapy should continue to govern the design of near term trials in gene therapy. Even if a definitive answer about the efficacy of the therapy cannot result from the study, the toxicity, immunogenicity, marking efficiency and stability of the vector system used can be ascertained. More recent examples of the importance of attention to these issues have come from studies in which adenovirus vectors are used to introduce genes into specific tissues. Rather than hope for unlikely clinical efficacy, attention is being paid to spread of virus throughout the body, immunologic response, extent and duration of expression, etc. In the long run, these data will prove to be very useful.

Are there any clinical trials of gene therapy likely to result in therapeutic benefit in the near future? Given the primitive nature of current vector systems, lack of knowledge about factors which determine stability of expression, and uncertain targeting, long-term therapy of human genetic disease or even therapy of somatic disorders seems unlikely soon. However, shorter term goals may well be achievable as efficiency of vector delivery, potency of promoters, and more careful choice of target tissues improves. Although not meant to be exhaustive, several illustrative areas of possible success in the near term are outlined below:

1. In cancer therapy, local regression of tumors, perhaps allowing surgical removal of tumors previously unamenable to such therapy, may prove possible. Several groups have begun to introduce tumor suppressor genes, such as p53, into adenoviral vectors (Roth and Cristiano 1997). Although spread of these defective viruses within tissues is limited by poor diffusion of large particles (and the use of fully replication competent viruses is not yet feasible due to safety concerns), preliminary results suggest that reduction of tumor mass may be possible. The use of chemotherapy for cytoreduction of breast cancers prior to surgery has recently been highlighted (Bonadonna et al. 1998). Some tumors which do not respond to chemotherapy may respond to gene therapy. Cure of cancer, especially metastatic disease, using gene therapy to directly target tumors or indirectly, via activation of the immune system, seems in the more distant future.
2. The use of gene therapy to improve tolerance to anti-cancer chemotherapeutic regimens, at least in the short term, seems within reach. It is possible to express drug-resistance genes, such as the multidrug transporter (*MDR*1), which confers resistance to anti-cancer drugs such as doxorubicin, vinblastine, or taxol (Gottesman et al. 1995), or a methotrexate-resistant dihydrofolate reductase (*dhfr*) (Zhao et al. 1997; Flasshove et al. 1998), in human bone marrow cells. Although the stability of this expression and the ability to target bone marrow stem cells may be problematic at this time, short-term expression in committed lineages may be sufficient to reduce toxicity during short courses of chemotherapy. Similarly, use of cy-

tokines, either delivered systemically as proteins or with gene therapy vectors, may allow dose escalation during chemotherapy of cancer.

3. Although many of the current gene therapy trials focus on the immune system, the likelihood of augmenting immune response in a way specific enough or potent enough to cure or alleviate cancer seems slim. As research on the immune response to cancer cells identifies better targets, however, some success may come in this area (Rosenberg 1997). The sheer magnitude of the resources and effort in this area may result in some breakthroughs in the next few years, although it is not clear at this time which approach is likely to work best.
4. Targeting angiogenesis and tumor blood vessels with specific genes encoding antiangiogenesis proteins seems feasible (Kong and Crystal 1998). The advantage of such an approach is that normal endothelial cells are not likely to become resistant to gene therapy because, unlike cancer cells, their mutation rates are low.
5. For treatment of monogenic inherited disorders, true long-term correction of defects in target cells is not likely to occur in the near future. However, delivery of vectors expressing gene products, either to defective cells or to other tissues, may allow short-term correction of metabolic abnormalities associated with genetic disease. One example which seems particularly promising is chronic granulomatous disease, in which only a small percentage of neutrophils need be corrected, and expression of NADPH oxidase complex in transduced cells need not be at wild-type levels (Malech et al. 1997). Patients are relatively well between infections, and short-term correction of neutrophils may help tide them over during an infection. Alternately, episodic treatment might prevent severe episodes of infection. Another approach is to express a gene product in a tissue from which it can be secreted, in cases in which the secreted product can act on a target cell (such as in lysosomal storage diseases; Brady and Barton 1996), or metabolize a circulating toxic product.
6. Treatment of cardiovascular disease using gene therapy shows conspicuous promise. Since short-term expression of factors which promote blood vessel growth (such as bFGF and VEGF) may allow development of collateral vasculature in a heart compromised by

coronary artery disease, efforts to develop this technology are already underway. Similarly, a major problem encountered after angioplasty is proliferation of smooth muscle in the vessels of coronary arteries whose endothelial cells are damaged by the procedure. A short-term gene therapy aimed at inhibiting this growth response, either by blocking release of trophic factors by the endothelium, or by blocking the mitotic response in smooth muscle, might prove effective. Another approach to cardiovascular disease is to engineer cells for specific purposes, such as endothelialization of artificial valves.

7. Although none of the early clinical trials of oligonucleotide therapeutics have shown efficacy, the promise of this approach is enormous, given the relative ease of delivery and the likelihood of specificity of anti-sense and other oligonucleotide therapeutics, such as introduction and correction of mutations using oligonucleotides which form triplex DNA (Wang et al. 1995, 1996). As the cost of synthesizing various oligonucleotide derivatives declines and the strategies for cellular uptake and synthesis of nonmetabolizable oligonucleotides improve, the likelihood of finding a suitable target for this therapy increases. Presumably, the best targets will be mutant genes in which correction of the genetic defect or reduction in expression of dominant negative genes does not have to be complete to give therapeutic benefit, or perhaps viral targets where a specific gene product may be limiting for viral growth.

1.4 Long-Term Goals

The imagination is the only real limit to a description of what gene therapy can accomplish in the more distant future. As a tool which can correct mutations, introduce new genes into cells, and regulate expression of existing genes, the possibilities for treatment of human disease are virtually boundless. Before such accomplishments are possible, however, certain technical barriers will need to be breached. These include the development of very high efficiency delivery systems for genes, cDNAs, and anti-sense DNAs and RNAs and the ability to target these sequences to specific tissues (either all tissues, or a subset of tissues needed to achieve a therapeutic effect). In many cases, homologous recombination between entering sequences and endogenous sequences will be necessary to guarantee substitution of new genes for existing genes, or stable expression of newly introduced genes. The ability to select for new genes, cells, or tissues which have received these genes will prove to be quite valuable, since new genes which are introduced may not have any specific selective advantage to the cells into which they are introduced. Finally, newly introduced genes should be regulatable, so that they may be turned on or off at will, and levels of their expression can be modulated.

1.4.1 Vectors

A great deal of attention is being directed towards development of safe, effective and efficient delivery systems. Viral vector systems have received a great deal of attention since some viruses are known to infect cells at high efficiency. Retroviruses are perhaps the best-studied vectors for gene therapy, and their potential utility remains substantial. However, retroviral promoters are not efficiently expressed in all cells, most retroviruses do not efficiently transduce nondividing cells, and retroviral titers are too low to produce enough particles to transduce large cell populations. These problems with retroviral vectors are not insurmountable, however. Other promoters have been introduced into these viruses; lentiviruses, of which HIV is a prominent example, do not require cell division for efficient transduction; and titers can be improved by use of transient packaging systems (allowing high level expression of what might be toxic packaging proteins) or by use of chimeric virus systems (Grignani et al. 1998). Targeting may also be improved by alterations in the envelope proteins of retroviruses (albeit, not easily). One problem with use of defective retroviruses in solid tissues or tumors is that the virus particles are large and do not easily diffuse through tissues. Use of replication- competent retroviruses, if they prove to be safe, might help solve this problem.

Other viral vector systems are also attractive and worthy of attention. Adeno-associated virus (AAV) is a DNA virus found in human populations which has no known pathogenicity. AAV-based vectors, once fully engineered, can be expected to introduce their genes into specific sites within the human genome, thereby reducing the risk of random, possibly carcinogenic, insertional events. Currently, AAV vectors have lost their native ability to insert at specific sites in the human genome and instead either insert randomly or exist as extrachromosomal segments (Baudard et al. 1996). Also, AAV is a defective virus which must be packaged using adenovirus as helper; there is a need to develop helper-free packaging systems for this virus and determine its tropism. Some recent experiments suggest that AAV may be an effective vector for gene therapy in mutant animals such as obese mice treated with an AAV-leptin vector (Murphy et al. 1997).

SV40 is another DNA virus whose utility for gene therapy is largely unexplored. It is able to encapsulate about 4 kb of DNA, was introduced into human populations during the early years of polio vaccination, and appears to cause no human disease. It infects many human cells at high efficiency. Two recent studies have suggested that it is possible to use SV40 to introduce marker and selectable (*MDR*1) genes into mouse and human hematopoietic tissues (Rund et al. 1998; Sandalon et al. 1997).

Herpes viruses are also attractive candidates for gene therapy, particularly in the nervous system (Fink and Glorioso 1997). These large complex viruses can be rendered nonpathogenic and have the capacity to carry large amounts of DNA into cells with high efficiency. Epstein-Barr (EB) virus introduces extrachromosomal DNA into cells (Robertson et al. 1996). If these replicating episomes can confer on their host cells a selective advantage, such as by use of drug-resistance selectable markers, then the danger of random integration into the genome can be eliminated in these systems. Since both EB virus and herpes virus are known human pathogens, and the exact details of their mechanisms of pathogenesis are not known, use of these vectors will require much more laboratory-based research.

Liposome-based systems have proven to be enormously flexible and surprisingly efficient in delivering DNA to cells in vivo (Gao and Huang 1995). Liposome-DNA complexes can be prepared in large amounts and relatively cheaply for therapeutic purposes. Targeting of liposomes is possible by use of surface antibodies or ligands. Although not yet practical for efficient gene transfer into humans, owing to trapping of liposomes in the reticuloendothelial system and in the first vascular bed which they encounter (usually the lungs if introduced intravenously), progress in the chemistry and biology of liposome-DNA preparations is substantial, and it is possible to imagine the use of liposomes for efficient gene delivery at some future time.

1.4.2 Targeting and Regulation

Targeting in the broadest sense refers to the ability to get a piece of DNA or RNA into a specific cell type or tissue and guarantee its expression in a specific target. Regulation is the ability to turn on or off the introduced sequences as needed for therapy, or to allow expression of the sequence only in the tissue of interest. Many different approaches are possible for each of these barriers to successful gene therapy. The use of vectors or liposomes with antibodies or ligands on their surface, either incorporated into existing envelope proteins or independently introduced chemically or genetically, may prove useful for targeting. Physical barriers to delivery, due to tissue turgor, barriers at the level of the capillaries or lymphatics, and diffusion through solid tumors and tissues, all compounded due to the large size of the particles involved in gene therapy, must be understood and conquered (Jain 1994). One approach, which exploits the biology of viruses, is to use replicating virus vectors and depend on cell to cell transmission of transducing virus, rather than physical spread of defective viruses. The possibility of this approach has been suggested by the difficulty of getting spread within human brain tumors of defective retrovirus producing cells carrying a herpes simplex virus thymidine kinase suicide gene (Ram et al. 1993).

Regulation can be achieved by use of promoters which are tissue-specific. Thus, delivery of a gene to a cell which can support transcription results in tissue-specific expression of that gene in the target tissue and not in tissues in which expression does not occur. Another similar approach is to use selectable markers (see below) in which toxicity is limited to a specific cell type or tissue, so that only in tissues sensitive to the selecting agent will transduced cells have a selective advantage and be enriched in the general population. An example is the use of the anti-cancer drug bisantrene, which is especially toxic to mouse B lymphocytes. When resistance to bisantrene is imparted by the multidrug-resistance gene (*MDR*1) introduced into bone marrow cells, the transgene is specifically enriched in B cells after bisantrene challenge (Aksentijevich et al. 1996).

The ability to turn expression of transgenes on and off once they have been introduced into cells is a much greater challenge. Clearly, homologous recombination, which inserts genes into their usual position in the genome, should allow normal regulation of transgenes. However, oftentimes it is desirable to turn on expression of genes in cells in which these genes are not normally expressed (such as, for example, expression of insulin or human growth hormone in transduced fibroblasts, as a therapy for diabetes, or growth factor deficiency). Ariad Pharmaceuticals has pioneered the use of engineered transcription units, which can be activated by drugs which directly cross-link components of the transcription apparatus to activate transcription (Amara et al. 1997). Approaches such as these show great promise in facilitating in vivo regulation of transgenes.

1.4.3 Stabilizing Expression of Transferred Sequences

Nature is very parsimonious and eventually turns off expression or actually deletes exogenous DNA sequences which are introduced into cells but do not confer upon those cells specific selective advantage. One obvious way to circumvent this universal problem in gene transfer is to improve homologous recombination of new sequences so that they simply replace nonfunctioning or damaged DNA sequences in recipient cells. Understanding the machinery involved in homologous recombination will be required in order to do this, and much progress remains to be made in this important area of basic biology. However, sometimes the replacement of homologous sequences is not possible, such as when the recipient cell does not contain these sequences, or not desirable, such as when independent regulation of a gene is necessary.

The traditional approach to this problem in genetics is the use of selectable markers. All geneticists and molecular biologists are familiar with the use of antibiotic resistance elements, such as ampicillin resistance and tetracycline resistance, to maintain the selective advantage of plasmids in bacteria, and the use of specific components of metabolic pathways to maintain genetic elements in auxotrophic yeast and bacteria. The principle is to improve the selective advantage of the recipient of a DNA segment based on information carried on that segment, and thereby enrich for transduced cells in a population. This principle can also be applied to gene therapy in animals and humans by use of drug-resistance markers, such as the *MDR*1 gene, which confers resistance to many different hydrophobic, cytotoxic drugs, and dihydrofolate reductase, whose expression can confer resistance to methotrexate and other inhibitors of this enzyme (Gottesman et al. 1995; Flasshove et al. 1998).

Transgenic mice in which the *MDR*1 gene is expressed in bone marrow are resistant to the leukopenic effect of many anti-cancer drugs (Galski et al. 1989), and transfer of the human *MDR*1 gene into hematopoietic cells of mice using a retroviral vector gives selective advantage to these cells after challenge with taxol (Sorrentino et al. 1992). Based on these studies, we have begun the design of bicistronic vectors in which an *MDR*1 cDNA is expressed on the same mRNA as a therapeutic, unselected gene separated by an internal ribosome entry site (IRES) to allow independent translation of both cDNAs. These vectors virtually guarantee that selectable multidrug-resistance is linked to co-expression of the nonselected gene of interest (Gottesman et al. 1995). Any tissue with rapid proliferation kinetics (bone marrow, skin, GI tract) can be the target for gene therapy with selectable markers. In addition to using these vectors to introduce cDNAs which encode nonselected therapeutic genes, the bicistronic mRNAs may also encode ribozymes whose high level selection is possible with the *MDR*1 gene (Lee et al. 1997).

1.5 Conclusions: A Bright Future

The intellectual concepts which form the basis for human gene therapy are undiminished by the practical obstacles which have become all too apparent during the rush to prove these concepts in clinical trials. The idea that a defective gene can be replaced or repaired and that new genes can be introduced to serve therapeutic goals is a powerful one whether or not this turns out to be easy to do. There is no flaw in the basic reasoning which led to the excitement about gene therapy. However, the translation of a concept to clinical practice is, and has always been, very difficult.

The more incremental approach to gene therapy which is currently the norm, rather than the exception, is likely to yield benefits in the long run. Carefully designed clinical trials will be useful tools to determine the toxicity and practicality of a myriad of new approach to gene delivery. We have no lack of ideas or the will to translate these into reality. It is just a matter of time before a gene therapy goes from the drawing board to practical application. We must be careful and patient. The future is bright.

References

Aksentijevich I, Cardarelli CO, Pastan I, Gottesman MM (1996) Retroviral transfer of the human MDR1 gene confers resistance to bisantrene-specific hematotoxicity. Clin Cancer Res 2:973–980

Amara JF, Clackson T, Rivera VM, Guo T, Keenan T, Natesan S, Pollock R, Yang W, Courage NL, Holt DA, Gilman M (1997) A versatile synthetic dimerizer for the regulation of protein-protein interactions. Proc Natl Acad Sci USA 94:10618–10623

Anderson WF (1995) Gene therapy. Sci Am 273:124–128

Baudard M, Flotte TR, Aran JM, Thierry AR, Pastan I, Gottesman MM (1996) Expression of the human multidrug resistance and glucocerebrosidase cDNAs from adeno-associated vectors: efficient promoter activity of AAV sequences and in vivo delivery via liposomes. Hum Gene Ther 7:1309–1322

Blaese RM, Culver KW, Miller AD, Carter CS, Fleisher T, Clerici M, Shearer G, Chang L, Chiang Y, Tolstoshev P, Greenblatt JJ, Rosenberg SA, Klein H, Berger M, Mullen CA, Ramsey WJ, Muul L. Morgan RA, Anderson WF (1995) T lymplocyte-directed gene therapy for ADA-SCID: initial trial results after 4 years. Science 270:475–480

Bonadonna G, Valagussa P, Brambilla C, Ferrari L, Moliterni A, Terenziani M, Zambetti M (1998) Primary chemotherapy in operable breast cancer: eight-year experience at the Milan Cancer Institute. J Clin Oncol 16:93–100

Brady RO, Barton NW (1996) Enzyme replacement and gene therapy for Gaucher's disease. Lipids 31 [Suppl]:S137–S139

Brenner MK (1996) Gene transfer to hematopoietic cells. N Engl J Med 335:337–339

Fink DJ, Glorioso J (1997) Engineering herpes simplex virus vectors for gene transfer to neurons. Nat Med 3:357–359

Flasshove M, Banerjee D, Leonard JP, Mineishi S, Li M-X, Bertino JR, Moore MAS (1998) Retroviral transduction of human $CD34^+$ umbilical cord blood progenator cells with a mutated dihydrofolate reductase cDNA. Hum Gene Ther 9:63–71

Galski H, Sullivan M, Willingham MC, Chin K-V, Gottesman MM, Pastan I, Merlino GT (1989) Expression of a human multidrug-resistance cDNA (MDR1) in the bone marrow of transgenic mice: resistance to daunomycin-induced leukopenia. Mol Cell Biol 9:4357–4363

Gao X, Huang L (1995) Cationic liposome-mediated gene transfer. Gene Ther 2:710–722

Gottesman MM, Hrycyna CA, Schoenlein PV, Germann UA, Pastan I (1995) Genetic analysis of the multidrug transporter. Annu Rev Genet 29:607–649

Grignani F, Kinsella T, Mencarelli A, Valtieri M, Riganelli D, Grignani F, Lanfrancone L, Peschle C, Nolan GP, Pelicci PG (1998) High-efficiency gene transfer and selection of human hematopoietic progenitor cells with a hybrid EBV/retroviral vector expressing the green fluorescence protein. Cancer Res 58:14–19

Jain RK (1994) Barriers to drug delivery in solid tumors. Sci Am 271:58–65

Kong H-L, Crystal RG (1998) Gene therapy strategies for tumor antiangiogenesis. J Natl Cancer Inst 90:273–286

Lee CGL, Jeang K-T, Martin MA, Pastan I, Gottesman MM (1997) Efficient long-term co-expression of a hammerhead ribozyme tartgeted to the U5 region of HIV-1 LTR by linkage to the multidrug-resistance gene. Antisense Nucl Acid Drug Dev 7:511–522

Malech HL, Maples PB, Whiting-Theobald N, Linton GF, Sekhsaria S, Vowells SJ, Li F, Miller JA, DeCarlo E, Holland SM, Leitman SF, Carter CS, Butz RE, Read EJ, Fleisher TA, Schneiderman RD, Van Epps DE, Spratt SK, Maack CA, Rokovich JA, Cohen LK, Gallin JI (1997) Prolonged production of NADPH oxidase-corrected granulocytes after gene therapy of chronic granulomatous disease. Proc Natl Acad Sci USA 94:12133–12138

Murphy JE, Zhou S, Giese K, Williams LT, Escobedo JA, Dwarki VJ (1997) Long-term correction of obesity and diabetes in genetically obese mice by a single intramuscular injection of recombinant adeno-associated virus encoding mouse leptin. Proc Natl Acad Sci USA 94:13921–13926

Orkin SH, Motulsky AG, co-chairs (1995) Report and recommendations of the panel to access the NIH investment in research on gene therapy. National Institutes of Health, Bethesda MD, http://www.nih.gov/news/panelrep.html

Ram Z, Culver KW, Walbridge S, Blaese RM, Oldfield EH (1993) In situ retroviral-mediated gene transfer for the treatment of brain tumors in rats. Cancer Res 53:83–88

Robertson ES, Ooka T, Kieff ED (1996) Epstein-Barr virus vectors for gene delivery to B lymphocytes. Proc Natl Acad Sci USA 93:11334–11340

Rosenberg SA (1997) Cancer vaccines based on the identification of genes encoding cancer regression antigens. Immunol Today 18:175–182

Roth JA, Cristiano RJ (1997) Gene therapy for cancer: what have we done and where are we going? J Natl Cancer Inst 89:21–39

Rund D, Dagan M, Dalyot N, Kimchi-Sarfaty C, Schoenlein P, Gottesman MM, Oppenheim A (1998) Efficient transduction of human hematopoietic cells with the human multidrug resistance gene 1 (MDR1) via SV40 pseudovirions. Hum Gene Ther 9:649–657

Sandalon Z, Dalyot-Herman H, Oppenheim AB, Oppenheim A (1997) In vitro assembly of SV40 virions and pseudovirions: vector development for gene therapy. Hum Gene Ther 8:843–849

Sidransky E, Fartasch M, Lee RE, Metlay LA, Abella S, Zimran A, Gao W, Elias PM, Ginns EI, Holleran WM (1996) Epidermal abnormalities may distinguish type 2 from type 1 and type 3 of Gaucher disease. Pediatr Res 39:134–141

Sorrentino BP, Brandt SJ, Bodine D, Gottesman MM, Pastan I, Cline A, Nienhuis AW (1992) Retroviral transfer of the human MDR1 gene permits selection of drug resistant bone marrow cells in vivo. Science 257:99–103

Wang G, Levy DD, Seidman MM, Glazer PM (1995) Targeted mutagenesis in mammalian cells mediated by intracellular triple helix formation. Mol Cell Biol 15:1759–1768

Wang G, Seidman MM, Glazer PM (1996) Mutagenesis in mammalian cells induced by triple helix formation and transcription-coupled repair. Science 271:802–805

Zhao RC, McIvor RS, Griffin JD, Verfaillie CM (1997) Gene therapy for chronic myelogenous leukemia (CML): a retroviral vector that renders hematopoietic progenitors, methotrexate-resistant and CML progenitors functionally normal and nontumorigenic in vivo. Blood 90:4687–4698

2 Approaches to the Gene Therapy of Cancer Using Replication-Competent Oncolytic Adenoviruses

C. Heise, D.H. Kirn, and A. Balmain

2.1 Introduction

A great deal has already been written about the potential for gene therapy in the development of novel approaches to the treatment of human cancer. The possibility of developing highly specific therapies based on the genetic alterations which take place during human tumor development seems very attractive, and in spite of the obvious technical hurdles which remain to be overcome, seems to be well worth substantial investment. Although some advances have been made in the more traditional approaches to cancer treatment, including novel surgical methods, radiation and chemotherapy, the fact remains that the overall survival in cancer patients has not improved dramatically over the past 40 years. This lack of improvement in therapeutic outcome is in stark contrast to the increase in our knowledge of the genetic events that are

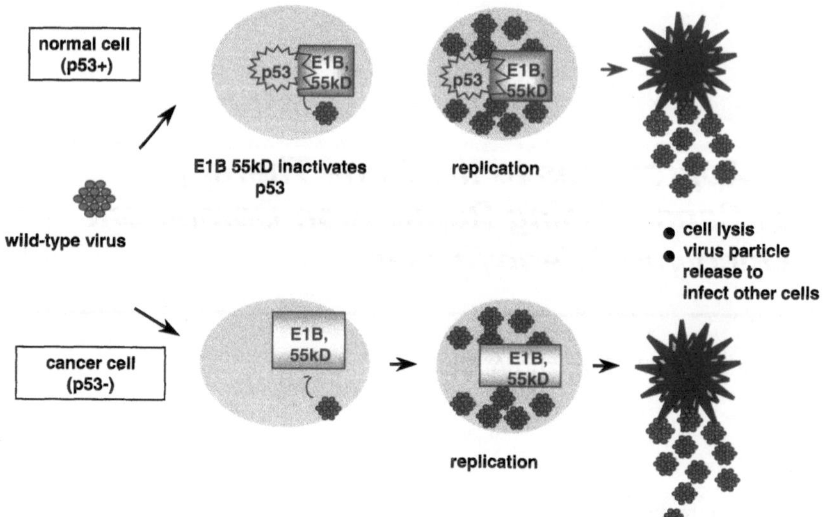

Fig. 1A, B. A In wild-type adenovirus infection, p53 is up-regulated following production of E1A gene products. This would normally result in abrogation of viral replication through cell cycle arrest or apoptosis. However, the E1B-55 kDa protein can bind and inactivate p53, thereby allowing completion of virus replication. **B** ONYX-015 adenovirus contains a deletion in the E1B-55 kDa gene, therefore p53 is not inhibited. This abrogates replication in normal cells; however, cancer cells that lack functional p53 are susceptible to replication and lysis by ONYX-015

involved in the genesis of many human tumors. Studies on the molecular genetics of cancer have led to the identification of a relatively large number of oncogenes and tumor suppressor genes, which suffer somatic mutations resulting in gain or loss of function, respectively, during tumor initiation and progression (Balmain 1997). Much emphasis is now being placed on the translation of this knowledge of cancer genetics into practical benefits in terms of novel therapy.

A number of companies have now introduced into clinical trials a series of potential therapeutic agents based on specific targets known to be activated in tumors. These targets generally are oncogenes which have acquired gain of function mutations or increases in activity at different stages of tumor development. The therapeutic approaches include the introduction of anti-sense oligonucleotides, antibodies which

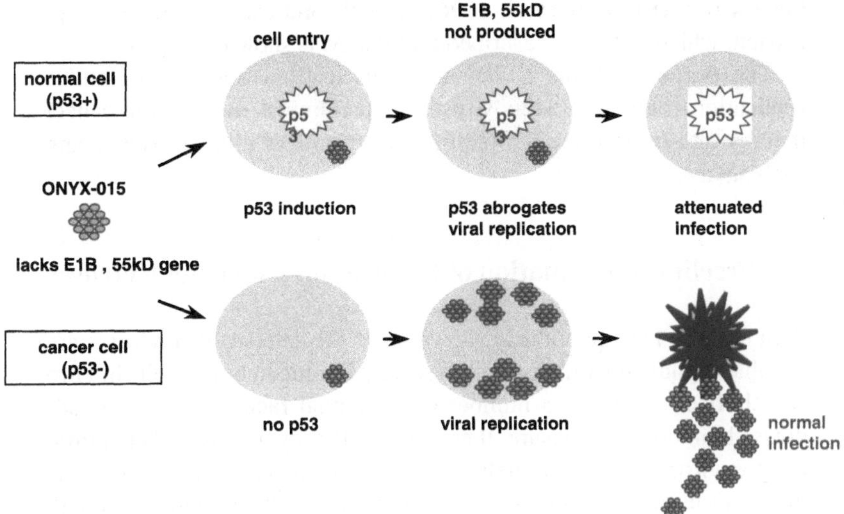

Fig. 1B. Legend see p. 18

bind and inactivate the appropriate target, small molecule inhibitors of specific tyrosine kinases (for example, PDGF or EGF receptors), inhibitors of ras signaling (farnesyl transferase inhibitors) or inhibitors of positive regulators of cell cycle progression (cyclin-dependent kinases or cdks) (Baringa 1997). All these approaches involve inhibition of the functions of activated oncogene activity within tumors. However, the most frequent genetic events within human tumors involve inactivation of tumor suppressor genes such as p53 or p16/INK4. Targeting the loss of function of tumor suppressor genes using small molecule approaches is conceptually much more difficult than targeting of activated oncogenes. A number of groups have therefore adopted gene therapy approaches in an attempt to introduce the functional tumor suppressor genes back into tumor cells. One such approach is exemplified by the attempt to introduce a wild-type p53 gene in an adenovirus vector into human lung tumors (Roth et al. 1996). An alternative approach, which has been exploited by Onyx Pharmaceuticals, is based on not p53 gene therapy, but rather an attempt to exploit the absence of p53 in many tumors to kill the cells using a replication-competent adenovirus,

ONYX-015 (Bischoff et al. 1996). The theoretical basis for this approach, which has been described previously, is shown in Fig. 1.

The purpose of this review is to discuss our attempts to utilize replication-competent adenoviruses in preclinical models for cancer therapy and to describe the preliminary results of clinical trials using such agents.

2.2 Preclinical Evaluation of Replication-Competent Viruses

Although we have previously shown that E1B-deleted adenoviruses are capable of inducing efficient cell lysis of p53 mutant tumor cells in vitro (Bischoff et al. 1996), a number of additional factors need to be addressed in order to ensure therapeutic efficacy in vivo. The tumor microenvironment is extremely complex and much more heterogeneous than cells used for in vitro assays. Tumors are frequently infiltrated with normal stromal or immune cells. The vasculature is highly disorganized and the tumors can also be fibrotic. A number of parameters have been investigated to optimize the delivery of therapeutic viruses to tumors growing as subcutaneous xenografts in nude mice. The results have shown that efficacy of intratumoral injections is greatly improved by increasing the frequency of applications of the virus while maintaining the same overall dose (Heise et al., submitted for publication). For example, a single injection of 5×10^8 plaque-forming units (PFU) was significantly less effective in inhibiting tumor growth than five daily injections of 1×10^8 PFU. Similarly, the efficacy was substantially improved by increasing the volume of medium in which the virus was suspended. Suspension of the virus in the higher volume before injection intratumorally seemed to have the effect of facilitating distribution of the virus to the extremities of the tumor rather than limiting localization to the site of injection. The results of these studies have therefore shown that it is possible to attain significant tumor inhibition and in some cases complete regression of subcutaneously growing human xenografts after five daily injections of this new therapeutic agent.

2.3 Approaches to the Systemic Administration of Oncolytic Viruses

In previous studies, it was demonstrated that direct injection of tumors in nude mice can, in some cases, result in a productive infection, which leads to dissemination of the virus and entry into subcutaneously growing tumors on the other flank, which had not been directly injected (Heise et al. 1997c). This demonstrated that viruses are capable of reaching distant tumor sites through the bloodstream and opened up the possibility of developing a systemic administration protocol for virus treatment. The possible problems involved in such an approach include the potential toxicity due to the uptake of adenoviruses by the liver, which is well known. We therefore carried out a series of experiments to assess the levels of ONYX-015 in subcutaneous tumors and in the liver after intravenous administration to nude mice. Preliminary results demonstrate that at 3 h and 6 h after virus administration, most of the detectable virus is indeed found in the liver, with a much lower amount in the subcutaneous tumors (Heise et al., unpublished). However, at 24–72 h after virus treatment, the concentration of virus found in the liver decreases substantially, whereas that found in the tumor increases, presumably owing to efficient replication within the human tumor cells. The net result is that there is a highly selective enrichment of the oncolytic adenovirus in the subcutaneously growing tumor cells. Using C33A cervical tumor cells as the target, this intravenous administration approach has been shown to significantly inhibit the growth of tumors (Heise et al. 1997a). The inhibition is clearly associated with the presence of replication-competent virus within the tumors. Immunohistochemistry for adenovirus hexon protein demonstrated the presence of replicating adenovirus adjacent to areas of vascularization in both C33A and HCT116 human tumor xenografts.

In conclusion, we have demonstrated that systemic administration of an oncolytic virus is feasible, that it can result in tumor growth inhibition due to virus replication, and that the amount of virus detectable in the liver decreases very substantially after 24 h. The limitations of this model are, of course, that these experiments are carried out in nude mice, which are immunocompromised and in which the normal cells are largely nonpermissive for replication of a human adenovirus.

2.4 Combination Therapies

There are substantial reasons for believing that combination of treatment with an oncolytic adenovirus with some known chemotherapeutic agents may have beneficial effects in tumor therapy. These agents have different mechanisms of action and presumably nonoverlapping toxicity profiles. In addition, it is well known that the adenovirus E1A gene can sensitize cells to the effects of chemotherapeutic agents or radiation (Lowe et al. 1993; Sanchez-Prieto et al. 1995; Brader et al. 1997). Such effects could synergize with the replication- mediated tumor cell lysis to induce a substantial increase in the therapeutic index. A series of studies have been carried out to test the efficacy of ONYX-015 in combination with a range of standard chemotherapeutic agents for tumors of the head and neck, ovary, colon and pancreas. Our results show that treatment of HLaC head and neck tumor cells with either ONYX-015 alone or 5-fluorouracil (5-FU) alone has less effects on cell growth than combinations of these two agents, which leads to substantial cell killing (Heise et al. 1997b, 1998). Similar results were obtained using the HCT-116 colon carcinoma cell line. These promising in vitro studies have led us to conduct a series of preclinical in vivo experiments, which have demonstrated the efficacy of these combination treatments. Our results clearly show that combinations of ONYX-015 with several chemotherapeutic agents were more effective than either agent alone, both in in vitro assays and in in vivo treatment of human tumor xenografts. Promising results were obtained, both by intratumoral injection and by intravenous administration of the viruses (Heise et al. 1997a). These results suggest that clinical trials of combined virus-chemotherapy treatments are warranted in human cancer patients. Such clinical trials are underway.

2.5 Phase I Trial of Onyx-015 in Treatment of Cancers of the Head and Neck

On the basis of the extremely promising results seen in in vitro assays and in several preclinical models, a Phase I trial was initiated to test the safety and efficacy of ONYX-015 alone in intratumoral injections into tumors to the head and neck (Kirn et al. 1997; Ganley et al. 1997). Head and neck tumors were chosen for the study because of the easy evaluation of the patients and the accessibility of the tumors for injection and biopsy. In addition, it is known that recurrent tumors of the head and neck have an extremely high incidence of p53 abnormalities, suggesting that they would constitute an efficient target for p53-dependent oncolytic adenovirus therapy. The patients chosen for the Phase I trial had recurrent disease which was refractory to chemotherapy and/or radiation therapy. To date, a series of 32 patients have been treated, of which 11 received multiple doses of the agent. No dose-limiting toxicity was observed, although in some cases mild and transient flu-like symptoms were detected in 30%–60% of the patients in different dose cohorts. The relative lack of toxicity in these studies is emphasized by the fact that in more recent treatment protocols normal cells in some patients have clearly been exposed to relatively high concentrations of virus without showing any obvious necrosis or local toxicity to the agent. It has also been possible to demonstrate in these studies that, although some of the patients have preexisting neutralizing antibodies to adenovirus, this does not preclude efficacy of the agent upon first administration.

In the preclinical subcutaneous tumor xenograft models, replication of ONYX-015 could clearly be detected in the regressing tumors. In primary human tumor biopsies, the level of detection was substantially lower than that seen in the nude mouse models, but was nevertheless detected in many patients, particularly those who had been injected multiple times with the virus. It could clearly be shown that the virus was present in cells adjacent to some necrotic areas and that the virus was replicating and inducing cellular degeneration. Electron microscopy demonstrated crystalline arrays of replicating adenovirus within the tumor biopsies, a clear indication that the normal lytic cycle of replication and lysis was operating within these primary tumors.

The overall results from this Phase I trial are that multiday dosing is clearly superior to the single injection protocol, as was previously predicted from the preclinical model. Responses could clearly be seen in patients in spite of the presence of neutralizing antibodies, and in those situations in which tumor necrosis was observed, this could, in some cases, be associated with clinical benefit in terms of improved swallowing, speech, etc. One additional observation made in these studies was that the response in individual patients did not vary as a direct function of the p53 sequence in the patients' tumors. This was predicted from previous in vitro studies where it was shown that many human tumors which have wild-type p53 sequence were nevertheless responsive to the ONYX-015 adenovirus. The reasons for this are not completely clear but it is known that there are many avenues by which the function of p53 can be abrogated in addition to mutation of the protein coding sequence. Indeed, studies using several cell lines with wild-type p53 gene sequences which are sensitive to ONYX-015 have shown an impaired or defective radiation-induced, p53-mediated G1 arrest, indicating that the pathway has been abrogated by a genetic change other than an intragenic p53 mutation. Although approximately half of all cancer patients have tumors that exhibit p53 mutations, this result shows that potentially a far greater proportion of all human tumors may be amenable to this type of oncolytic therapy owing to functional inactivation of p53 by other mechanisms.

In conclusion, these studies represent a new paradigm for the exploitation of loss of a tumor suppressor gene for cancer therapy. One could envisage that the efficacy of an agent such as ONYX-015 may be substantially improved by further engineering of the virus itself, by the introduction of additional mutations which may enhance replication, lysis and spread within tumors, or by incorporation of exogenous genes which may lead to an enhanced immune response or improved tumor cell killing. Although the introduction of such genes into human tumors is presently being attempted using replication-deficient gene therapy approaches, the use of a replication-competent vector should offer significant advantages in terms of gene delivery and the overall magnitude of the response (Kirn 1996; Kirn and McCormick 1996).

References

Balmain A (1997) Tumour suppressor genes as potential targets for the action of carcinogens. Compr Toxicol 12:83–110

Baringa M (1997) From bench top to bedside. Science 278:1036–1040

Bischoff J, Kirn D, Williams A, Heise C, Horn S, Muna M, Ng L, Nye J, Sampson-Johannes A, Fattaey A, McCormick F (1996) An adenovirus mutant that replicates selectively in p53- deficient human tumor cells. Science 274:373–376

Brader KR, Wolf JK, Hung M, Yu D, Crispens M, van Golen K, Price J (1997) Adenovirus E1A expression enhances the sensitivity of an ovarian cancer cell line to multiple cytotoxic agents through an apoptotic mechanism. Clin Cancer Res 3:2017–2024

Ganly I, Kirn D, Rodriquez GI, Soutar D, Eckhardt G, Otto R, Robertson AG, Park O, Gulley ML, Kraynak M, Heise C, Maack C, Trown PW, Kaye S, Von Hoff D (1997) Phase I trial of intratumoral injection with an E1B-attenuated adenovirus, ONYX-015 in patients with recurrent p53(-) head and neck cancer. Proc Am Soc Clin Oncol 16:382a

Heise C, Sampson-Johannes A, Williams A, McCormick F, Von Hoff D, Kirn D (1997a) ONYX-015, an E1B gene-attenuated adenovirus, causes tumor-specific cytolysis and antitumoral efficacy that can be augmented by standard chemotherapeutic agents. Nat Med 3:639–645

Heise C, Williams A, Propst M, Sampson-Johannes A, Weber S, Davidson K, Izbicka E, VonHoff D, Kirn D (1997b) Preclinical studies with ONYX-015 (a replication competent E1B-deleted adenovirus) in combination with chemotherapy. Cancer Gene Ther 4:S13

Heise C, Propst M, Williams A, Mangold G, Von Hoff D, Kirn D (1997c) Antitumor efficacy following intravenous administration of a replication-competent, attenuated adenovirus in nude mice. Proc Am Assoc Cancer Res 38:10

Heise C, Williams A, Xue S, Davidson K, Izbicka E, Mangold G, Von Hoff D, Kirn D (1998) Combination therapy studies with Onyx-015 (a replication competent E1B-deleted adenovirus) and chemotherapeutics. Proc Am Assoc Cancer Res 39:419

Kirn DH (1996) Replicating oncolytic viruses: an overview. Exp Opin Invest Drugs 5: 753–762

Kirn DH, McCormick F (1996) Replication viruses as selective cancer therapeutics. Mol Med Today 519–527

Kirn D, Ganley I, Nemunaitis J, Otto R, Soutar D, Kuhn J, Heise C, Propst M, Maack C, Eckhardt G, Kaye S, VonHoff D (1997) A phase I clinical trial with ONYX-15 (a selectively replicating adenovirus) administered by intratumoral injection in patients with recurrent head and neck carcinoma. Cancer Gene Therapy 4:S13

Lowe SW, Ruley HE, Jacks T, Housman DE (1993) p53-dependent apoptosis modulates the cytotoxicity of anticancer agents. Cell 74:957–967

Roth JA, Nguyen D, Lawrence DD, Kemp BL, Carrasco CH, Ferson DZ et al (1996) Retrovirus-mediated wild-type p53 gene transfer to tumors of patients with lung cancer. Nat Med 2:985–991

Sanchez-Prieto R, Lleonart M, Cajal R (1995) Lack of correlation between p53 protein level and sensitivity to DNA-damaging agents in keratinocytes carrying adenovirus E1A mutants. Oncogene 11:675–682

3 Discovery, Development, and Application of Synthetic Gene Delivery Systems

P.L. Felgner

3.1 Introduction

More than 25 years ago, in a *Science* article, Friedmann and Roblin outlined prospects for human gene therapy (Friedmann and Roblin 1972). This forward-looking review anticipated the development of two alternative gene delivery systems: viral gene therapy vectors and synthetic gene delivery systems using purified gene sequences. Theoretical support for the use of viruses arose from the knowledge that DNA and RNA tumor viruses were capable of introducing heritable changes into the genome of mammalian cells. Friedmann proposed that, if the deleterious features of these viruses could be eliminated, one might safely

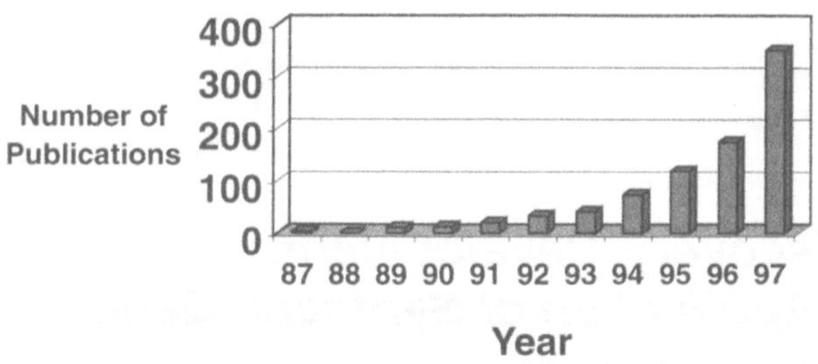

Fig. 1. Annual publication rate in synthetic gene delivery system technology

deliver a desirable gene to correct cellular defects or genetic disorders. As molecular biology techniques matured, the tools to package genes into nonreplicating, recombinant retroviral vectors became available (Mann et al. 1983). These vectors allowed investigators to introduce recombinant genes into living cells and to permanently transduce them with a new genetic phenotype. During the last 15 years substantial progress has been made to produce safe and effective viral vectors, and we have witnessed an exponential growth in preclinical research and clinical development of recombinant viral vectors for gene therapy applications (Fig. 1).

The introduction of synthetic, nonviral gene delivery systems into the clinical gene therapy repertoire has taken somewhat longer to develop. Until very recently, most investigators active in gene therapy research considered it unlikely that nonviral vectors would be able to compete with the gene delivery efficiency of highly evolved viruses for gene therapy applications. Two discoveries that came out of my labs at Syntex and Vical significantly altered this point of view. The first demonstration was that cationic liposomes could be used very conveniently to efficiently transfect plasmids into cultured cells (Felgner et al. 1987). The levels of expression obtained were often comparable to those seen with recombinant viral vectors. These results were followed by several reports demonstrating that the cationic lipids could also enhance expression from plasmids administered directly in vivo. Our second major finding came out of studies designed to evaluate the in vivo transfection activity of cationic liposomes. These studies led to the surprising finding that a very significant level of in vivo expression could be obtained following intramuscular injection of naked plasmid

Table 1. Synthetic gene delivery systems

DEAE dextran
Calcium phosphate
Lipopolyplex – Cationic lipid-based
Polyplex – Polycationic polymer-based
Lipopolyplex – Mixture of lipids and polymers
Naked DNA
Electroporation
Gene gun

DNA (Wolff et al. 1990). We subsequently extended these findings and demonstrated that animals could be immunized by injecting plasmids encoding viral antigens into skeletal muscle, thus launching the field of DNA vaccines (Ulmer et al. 1993). Together these findings significantly raised the level of interest in nonviral gene delivery approaches for gene therapy applications.

Practical manufacturing and clinical development issues also stimulated interest in the nonviral gene delivery system research and development. The plasmids used for these products are chemically well defined, can be highly purified and can be produced in large quantities by conventional bacterial fermentation. The broad applicability, ease of manufacturing, potential cost effectiveness, and safety (i.e., no risk of infection; low immunogenicity) of this gene-based drug therapy approach offers competitive advantages for research and commercialization, compared to the use of viral vectors (Felgner and Rhodes 1991). For all of these reasons today the scientific research and clinical application of nonviral gene therapy products is rapidly increasing.

As the number of investigators and scientific papers on the topic of synthetic gene delivery systems has increased, the terminology that describes these systems has also expanded. In many cases different investigators used different terms to describe the same type of system. Recognizing this problem a committee of investigators active in this area met to define common terminology (Felgner et al. 1997). Cationic lipid mediated transfection was termed "lipofection" and the cationic lipid-DNA complexes which form when cationic liposomes are mixed with DNA were termed "lipoplexes." Similarly, transfection mediated by hydrophilic polycations such as polylysine or dendrimers was termed polyfection and the complexes that form when these polycations are mixed with DNA are termed "polyplexes." Most of the synthetic gene delivery systems under active investigation today are lipoplex, polyplex, lipopolyplex or naked DNA (Table 1).

3.2 Historical Overview

Although synthetic delivery systems entered the gene therapy repertoire somewhat later than viral vectors there is a long history of nonviral gene delivery research. In 1928 Fred Griffith was working with two different strains of the pneumococcus bacteria that exhibited different pathogenicities and colony morphologies. He showed that heat-killed pathogenic cells, when mixed with the live nonpathogenic cells, could transform a small percentage of the nonpathogenic cells into the pathogenic phenotype. Oswald Avery took up the task of defining the active principle responsible for this transformation. He originally believed that Griffith's "transforming factor" was a complex polysaccharide because the wild-type and transformed colonies had different morphologies, and he reasoned that this characteristic difference might have been due to differences in the bacterial cell surface carbohydrates (Avery et al. 1944). In 1944 he showed that the transforming factor could be extracted and purified with the nucleic acid fraction from the heat-killed bacteria. It was stable to protease and RNase digestion, but sensitive to DNase digestion. These results convinced most scientists that the active principle in transforming factor was DNA, and thus these experiments were among the first demonstrations of DNA transfection. Today bacterial cell transfection is a routine molecular biology technique that is essential to the application of recombinant DNA technology.

3.3 In Vitro Transfection in Mammalian Cells

The earliest efforts to identify methods for enhancing delivery of functional purified polynucleotides into living mammalian cells were stimulated in the mid-1950s by the results of Alexander and Holland (Alexander et al. 1958; Holland et al. 1959), showing that purified poliovirus

Table 2. Methods developed to deliver infectious viral nucleic acid in vitro

Year	Author	Milestone
1958	Alexander et al.	Purified infectious poliovirus RNA
1959	Holland et al.	Purified infectious poliovirus RNA
1962	Smull and Ludwig	Basic protein-mediated
1965	Vaheri and Pagano	DEAE dextran-mediated
1973	Graham and Van der Eb	Calcium phosphate method

RNA was infectious in HeLa cells (Table 2). Since the titer of this purified genomic RNA stock was extremely low (up to 1 million times less active than the original intact virus), it was questioned whether a vanishingly small quantity of intact virus had survived the purification procedure. They showed that a hypertonic saline solution (1 M NaCl) could enhance infectivity of the purified RNA by about 100-fold. This result, which demonstrated that living virus particles were not required to produce living (i.e., infectious and replicating) virus, was subsequently confirmed and extended, using improved methods for polynucleotide delivery including calcium phosphate coprecipitation (Graham and van der Eb 1973) and DEAE dextran (Vaheri and Pagano 1965). Broader utility for the calcium phosphate procedure was later demonstrated by experiments showing that transfection of a noninfectious fragment of the herpes simplex virus genome containing the thymidine kinase gene could transform thymidine kinase-negative cells (Minson et al. 1978). This result demonstrated that an infectious and proliferating virus was not necessary to induce cellular transformation leading to a new cellular phenotype. Transfection technology and the emerging recombinant DNA technology converged in the late 197's when Berg and colleagues applied calcium phosphate, DEAE dextran and liposome mediated transfection methodologies to the delivery and expression of recombinant plasmids in cultured mammalian cells (Table 3) (Mulligan et al. 1979; Fraley et al. 1980; Southern and Berg 1982). These findings have led to widespread applications for transfection methodology in molecular and cellular biology and in the pharmaceutical industry.

In the mid-1980s a transfection technology emerged from my laboratory that was based on the use of cationic liposomes (Table 4) (Felgner et al. 1987). Since at that time there were no chemically stable, liposome-forming cationic lipids available, we synthesized a series of

Table 3. Synthetic gene delivery system applications

Year	Author	Milestone
1978	Minson et al.	Thymidine kinase transformation
1979	Mulligen et al.	Rabbit β-globin plasmid
1982	Southern and Berg	Neomycin (G418) selection
1983	Mann et al.	Helper-free retrovirus vector
1984	Hwang and Gilboa	Published article "...retroviral infection is more efficient than...DNA transfection"

Table 4. Second generation synthetic gene delivery systems (systems developed to delivery plasmid in vitro and in vivo)

Year	Author	Milestone
1980	Fraley et al.	Liposomes
1987	Felgner et al.	Cationic lipids
1987	Wu and Wu	Polylysine/receptor mediated
1988	Johnston et al.	Gene gun

molecules that could form stable cationic liposomes. The prototype for this approach was a novel positively charged lipid, DOTMA (N[1-(2,3-dioleyloxy) propyl]-N,N,N-trimethylammonium) (Fig. 2). DOTMA forms liposomes that interact spontaneously with DNA or RNA, resulting in a liposome/polynucleotide complex, called a lipoplex. Lipoplexes are capable of delivering functional nucleic acid molecules into tissue culture cells. Efficacious cationic liposome formulations have three properties that are particularly important to polynucleotide delivery applications (Felgner and Ringhold 1989). First, these vesicles spontaneously condense with DNA to form a complex in which up to 100% of the DNA is entrapped. This high entrapment efficiency is not limited by the size of the DNA. Second, virtually all biological surfaces, including cultured cell surfaces, carry a net negative charge. Consequently, positively charged lipid vesicles, complexed with DNA or RNA, interact spontaneously with the negatively charged cell surfaces delivering the associated polynucleotide to the cell surface. And finally, cationic lipids fuse with cell membranes in a manner that allows the entrapped DNA to enter the cytoplasm and escape from the degradative lysosomal pathway. Although receptor-mediated endocytosis usually enables even con-

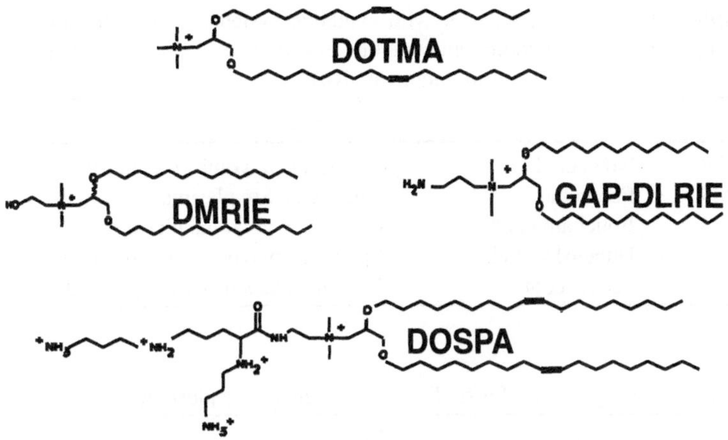

Fig. 2. Examples of cationic lipid molecules that are effective in transfection assays

ventional liposomes to enter cells in reasonable amounts, these particles do not easily escape the lysosome, where the liposomes and their encapsulated DNA can be completely degraded. Cationic lipid mediated fusion depends on the structure of the cationic lipid molecules as well as on the presence of neutral lipids in the final formulation. Today, many different cationic liposome formulations have been found to be effective for enhancing gene delivery and more than 30 products are commercially available for this purpose, including Lipofectin, TransfectAM, LipofectACE, LipofectAMINE, Superfect and even LipoTaxi (Fig. 2).

3.4 In Vivo Gene Delivery — Infectious and Oncogenic Systems

As discussed previously, the earliest successful in vitro transfection experiments were done with purified infectious viral genomes and the endpoint was the demonstration of infectious virus particles or plaques (Alexander et al. 1958; Holland et al. 1959). Similarly, among the earliest reported in vivo transfections were experiments showing that virus replication could occur following injection of purified or cloned

Table 5. In vivo gene transfer with synthetic systems using infectious and oncogenic plasmids (studies showing that infectious plasmids replicate in vivo and oncogenic plasmids form tumors in vivo)

Year	Author	Milestone
1979	Israel et al.	Infectious polyoma virus plasmid
1983	Fung et al.	Oncogenic Src plasmid
1984	Bouchard et al.	Oncogenic polyoma plasmid
1984	Dubensky et al.	Infectious polyoma virus plasmid
1984	Seeger et al.	Infectious hepatitis virus plasmid

Table 6. Early gene transfer studies with noninfectious plasmids

Year	Author	Milestone
1983	Nicolau et al.	Insulin gene lowers glucose levels Liposomes
1986	Benvenisty and Reshef	CAT plasmids in peritoneum Calcium phosphate
1987	Wang and Huang	CAT plasmid in peritoneal tumor liposomes
1989	Wu et al.	CAT plasmids in liver Polylysine/orosomucoid
1989	Kaneda et al.	Reporter plasmids in liver Liposomes
1989	Brigham et al.	Lung plasmids in lung Cationic lipids
1990	Nabel et al.	β-Galactosidase in blood vessel Cationic lipids
1990	Holt et al.	Luciferase in brain Cationic lipids
1990	Yang et al.	Reporter genes – liver, skin, muscle Gene gun
1990	Wolff et al.	Luciferase and CAT in muscle Naked DNA

viral genomes in vivo (Table 5), and the highest levels of expression observed were with particulate formulations of calcium phosphate-precipitated DNA (reviewed in Felgner and Rhodes 1991). Israel et al. (1979) showed that polyoma virus DNA was infectious following intraperitoneal injection into mice and hamsters, but at a level four to five logs below that of intact polyoma virions. Dubensky and Bouchard showed polyoma virus DNA replication efficiency could be improved by using calcium phosphate-precipitated DNA and by treating the tissue with hyaluronidase and collagenase (Table 6) (Dubensky et al. 1984; Bouchard et al. 1984). Both groups also showed that the infectious polyoma virus plasmid produced tumors in a significant percentage of the treated animals. Fung showed that a noninfectious plasmid encoding the *src* oncogene could produce tumors in suseptible chickens (Fung et al. 1983). Gould-Fogerite used liposomes to deliver a noninfectious, oncogenic papyloma virus plasmid DNA fragment to induce mouse tumors following direct subcutanteous injection (Gould-Fogerite et al. 1989). And finally, Varmus and coworkers gave a single 20 μg intrahepatic injection of a cloned infectious hepatitis viral DNA (ground squirrel specific virus) into ground squirrels and produced seropositive animals at 11–18 weeks postinjection (Seeger et al. 1984). In all of these cases the observed effects were primarily attributed to the ability of the systems under study to be amplified by replication of an extremely rare, low frequency infectious or oncogenic event. Discussion of how the apparent low level of expression obtained could be used to therapeutic benefit was limited. However, the results did show that infectious and oncogenic DNA sequences could be taken up and expressed by cells following direct in vivo injection.

3.5 In Vivo Gene Expression with Nonreplicating Systems

During the 1980s several reports were published showing in vivo expression from directly injected noninfectious, nononcogenic plasmid DNA sequences (Table 7). In vivo gene expression was reported using liposomes, calcium phosphate, a polylysine/protein conjugate, cationic lipids, and naked DNA. Benvenisty detected chloramphenicol acetyl transferase (CAT) activity in newborn rat tissues following intraperitoneal injection of calcium phosphate precipitated plasmid DNA (Ben-

Table 7. Third generation synthetic gene delivery systems (improved systems for in vitro and in vivo gene delivery)

Year	Author	Milestone
1990	Wagner et al.	Polylysine/receptor-mediated
1995	Boussif et al.	Polyethylenimine
1996	Tang et al.	Dendrimer

venisty and Reshef 1996). A panel of CAT plasmid constructs containing different eukaryotic viral promoters, a proinsulin gene and a human growth hormone gene were all reported to express messenger RNA and/or gene product.

Wu prepared an asialo-orosomucoid (AsOR)/polylysine conjugate which interacted spontaneously with pSV2 CAT plasmid DNA (Wu et al. 1989). Twenty-four hours following intraveneous injection, CAT activity was detected in rat liver. Activity was only detected when the pSV2 CAT DNA was complexed with the AsOR/PL conjugate; free plasmid DNA was not active in vivo. The activity declined to baseline levels by 96 h post-dosing. Animals given a partial hepatectomy prior to dosing maintained high levels of CAT activity for up to 11 weeks post-dosing. Analysis of DNA extracts from the tissues by Southern blot suggested that the CAT plasmid DNA was integrated into the genomic liver DNA. These results led to the hypothesis that cells undergoing division in regenerating liver permit stable integration of episomal DNA into the host cell genome and that this newly integrated DNA is active and stable.

The ability of liposomes to deliver genes in vivo was described in several reports. Nicholau reported a transient hypoglycemic effect and elevated insulin levels in the blood, spleen and liver of rats, following intravenous injection of about 2 μg of a plasmid containing the rat preproinsulin gene (Nicholau et al. 1983). Wang demonstrated CAT activity following intraperitoneal injection of pH sensitive immunoliposomes, containing a CAT plasmid, into nude mice bearing an ascites tumor (Wang and Huang 1987). Kaneda described experiments showing expression of SV40 large T antigen in the liver of rats following injection of a pBR SV40 plasmid into the portal vein (Kaneda et al. 1989). The delivery vehicle was a complex mixture of components mixed in a particular order so as to produce complexes of phospholipid vesicles

Table 8. Effective DNA vaccine immunization

Viruses	Avian and human influenza; bovine herpes; BVDV; Dengue fever; encephalitis; FELV; hepatitis B andC; herpes simplex, HIV-1; LCMV; measles; papilloma; rabies; RSV; SHIV, SIV, SV-40
Bacteria	Lyme; *Moraxella bovis*; *Mycobacterium tuberculosis*; *Mycoplasma*; *Rickettsia*; *Salmonella*; tetanus
Parasites	*Cryptosporidium parvum*; *Leishmania*; malaria; *Schistosoma*

containing encapsulated DNA, ganglioside, Sendai virus fusion proteins and red blood cell membrane proteins.

Cationic lipids were also studied for their ability to deliver genes in vivo. CAT gene expression was detected in the blood and lungs of mice following injection of 15 μg or 30 μg pSV2CAT plasmid complexed with cationic lipid vesicles (Lipofectin Reagent, BRL; Brigham et al. 1989). Injections were intraveneous, intraperitoneal and intratracheal. Expression was observed for 6 days post-injection. Holt showed that direct injection of CAT and luciferase DNA and RNA into *Xenopus leavis* embryos resulted in expression of luciferase gene product (Holt et al. 1990). Functional expression was shown to be dependent on the presence of the cationic lipid DOTMA, and expression could be enhanced by the coadministration of proteolytic enzymes. Luciferase activity in transfected embryos peaked during the first 48 h post-transfection and was still detectable 28 days later. And finally, Nabel showed that cationic lipids could enhance expression from plasmids administered into catheterized blood vessels (Nabel et al. 1989). Species in which DNA vaccine efficacy has been demonstrated include chicken, trout, mouse, rat, guinea pig, rabbit, ferret, cat, goat, sheep, cow, horse, monkey, pig, chimpanzee.

Interest in the nonviral gene delivery approach was boosted by our results showing in vivo reporter gene expression from plasmids encoding luciferase, CAT and β-galactosidase genes following direct injection into mouse muscle (Wolff et al. 1990). Up to 1 ng of gene product could be isolated from tissues injected with 10–100 μg of plasmid DNA and expression was shown to persist for more than 6 months. The DNA did not integrate into the host genome and plasmid essentially identical to the starting material could be recovered from the muscle months after injection. In vitro transcribed messenger RNA was also taken up and expressed in mouse muscle but the duration of expression was much shorter due to enzymatic breakdown of the message. Interestingly, no special delivery system was required for these effects. The practical potential of this nonviral approach to gene delivery was further substantiated by the demonstration that animals could be immunized following intramuscular injection of plasmids encoding heterologous antigens (Ulmer et al. 1993). This led to a rapidly expanding interest in the new field of DNA vaccines, which is being actively investigated by immunologists, clinicians and pharmaceutical companies today. Human clinical trials evaluating DNA vaccines for HIV, influenza, malaria and hepatitis are currently underway, and at least ten other vaccine candidates are in advanced preclinical development (Table 8).

3.6 Clinical Progress in Synthetic Gene Delivery Systems

The search for applications for synthetic gene delivery systems has spread throughout the major areas of clinical medicine (Table 9). DNA vaccines are being evaluated for the prevention and/or treatment of more than 20 different infectious diseases. At least four different approaches are being evaluated in human clinical trials for the treatment of cancer. And preclinical data is just now beginning to be generated suggesting that applications for metabolic disorders such as diabetes, cystic fibrosis and anemia may be treatable by gene therapy approaches utilizing synthetic delivery systems.

One of the most active areas of gene therapy clinical research is in oncology (Felgner et al. 1995; Nabel and Felgner 1993). It is possible to consider many different classes of genes, which if administered by a gene therapy approach either systemically, directly into tumor tissue, or

Table 9. Potential clinical applications

Infectious diseases	Two human clinical trials for malaria and influenza; 20 preclinical targets
Oncology	Four products in human clinical trials
Metabolic and genetic disorders	Human clinical trials (cystic fibrosis with CFTR; peripheral vascular disease with VEGF); preclinical investigation (EPO, leptin, insulin)

specifically into tumor cells, would exhibit a therapeutic or prophylactic anticancer effect. Three anticancer gene therapy products are in clinical development at Vical which utilize: (1) Interleukin (IL)-2 lymphokine therapy, (Leuvectin); (2) HLA-B7 heterologous antigen therapy (Allovectin-7); and (3) an anti-idiotype therapeutc DNA vaccine for B cell lymphoma (Vaxid).

Allovectin-7 is a gene-based lipoplex product intended for direct injection into tumor lesions of cancer patients. The product contains a gene that encodes a foreign tissue antigen (HLA-B7) which, when injected into tumors, is intended to cause the malignant cells to bear this antigen on their surface. When this foreign antigen is expressed, the patient's immune system, which previously failed to recognize the tumor cells as abnormal, may attack and destroy the cancer cells as if they were foreign tissue.

After a small pilot trial conducted by Nabel, Vical initiated Phase I/II clinical trials with approximately 15 patients for each of three advanced cancer indications: renal cell carcinoma, melanoma, and colorectal carcinoma (Nabel et al. 1995). The trials were designed primarily to test the safety of Allovectin-7 at varying dosage levels and to assess HLA-B7 gene transfer and expression. These studies showed that gene transfer was successful in the majority of patients, the treatment appeared to be safe and well tolerated, and measurable tumor shrinkage was observed in seven of 14 patients with advanced melanoma. A multicenter Phase II clinical trial of Allovectin-7 is currently underway in patients with six tumor types: melanoma, renal cell carcinoma, colorectal carcinoma, breast carcinoma, non-Hodgkin's lymphoma, and head and neck cancer. More than 100 patients have been treated so far. In addition, Allovectin-7 is being evaluated, either alone or in combination with approved cancer therapeutic agents, in several other Phase I/II clinical trials.

Leuvectin contains a gene that encodes the potent immunostimulator IL-2 and is also formulated as a lipoplex to facilitate gene uptake. When injected into tumors, Leuvectin causes the malignant cells to produce and secrete IL-2 in the vicinity of the tumor lesion. The local expression of IL-2 may stimulate the patient's immune system to attack and destroy the tumor cells. Recombinant IL-2 protein is a Food and Drug Administration (FDA)-approved anti-cancer agent for the treatment of advanced renal cell carcinoma. It has been investigated widely as a cancer immunotherapeutic agent, but is frequently associated with serious side effects. Because Vical's gene-based product candidate is designed to deliver IL-2 only at the site of tumor lesions, Leuvectin may provide similar efficacy with fewer side effects than systemic protein-based therapy. This is because the DNA, once introduced into the body, is intended to stimulate the production of a IL-2 at high local concentrations over a prolonged period of time. A major shortcoming of the existing recombinant IL-2 therapy is its short duration of action and the side effects associated with high levels of circulating protein after intravenous administration. By direct injection of the gene into the tumor, a sustained-release of the IL-2 within the tumor tissue may be achieved with fewer and less severe side effects.

The initial Phase I/II clinical trials were designed primarily to test the safety of Leuvectin at varying dosage levels and to assess IL-2 gene transfer and expression. The initial clinical results showed that gene transfer was effective in the majority of patients, the treatment appeared to be safe and well-tolerated, and measurable tumor shrinkage was observed in five of 23 patients with various types of advanced malignancies. An additional multicenter Phase I/II clinical testing of higher doses of Leuvectin is underway in approximately 45 patients with advanced melanoma, renal cell carcinoma, and soft-tissue sarcoma.

Ronald Levy and collaborators at Vical developed a naked DNA anti-idiotype vaccine (Hakim et al. 1996), Vaxid, against low-grade non-Hodgkin's B cell lymphoma. This type of lymphoma is characterized by a slow growth rate and excellent initial response to chemotherapy or radiotherapy; however, a regular pattern of relapse to a diffuse aggressive lymphoma occurs for which no curative therapy has been identified. Clinical studies involving administration of either monoclonal anti-idiotype antibodies or patient-specific B cell lymphoma idiotype protein have resulted in prolonged remissions; however, these therapies are limited by the time and effort required to produce the drug product. Vaxid is a DNA plasmid that encodes the patient-specific idiotype of the B cell tumor immunoglobulin. In preclinical studies, Dr. Levy showed that the injection into mice of a murine B cell lymphoma idiotype plasmid results in strong anti-idiotype immune responses and significant protection against tumor challenge. Based on these preclinical studies and additional studies conducted at Vical, the company believes that immunization of post-chemotherapy patients with Vaxid could result in the elimination of residual disease and the prevention of the relapse of disease. Vaxid is currently being tested in a human clinical trial.

3.7 Synthetic Gene Delivery System Efficiency and Efficacy

Although nonviral vectors are fundamentally attractive from a pharmaceutical development perspective, the efficacies of the relatively primitive delivery systems in use today are still relatively low. For example, viruses can infect cultured cells with nearly perfect efficiency, wherein 100 infectious virus particles containing 100 viral genomes can successfully infect almost 100 cells. In order to obtain similar levels of expression with nonviral vectors it typically takes 100 million copies of plasmid to transfect 100 cells.

A further illustration that nonviral gene therapy technology is currently in its infancy is shown by the following graph (Fig. 3). This figure compares the level of gene product expression obtainable today, after in vivo administration of a plasmid encoding a reporter gene, with the level that can be obtained in cultured cells. Here it can be seen that the amount of gene product recovered following intramuscular or intratumor plasmid injection is three to four orders of magnitude lower than that which can be obtained in cultured cells. Thus, there is room to improve in vivo nonviral gene delivery system technology before it reaches an efficiency level comparable to in vitro transfection. Scientists who succeed in pushing the in vivo expression levels up by one, two or three orders of magnitude will create new opportunities to

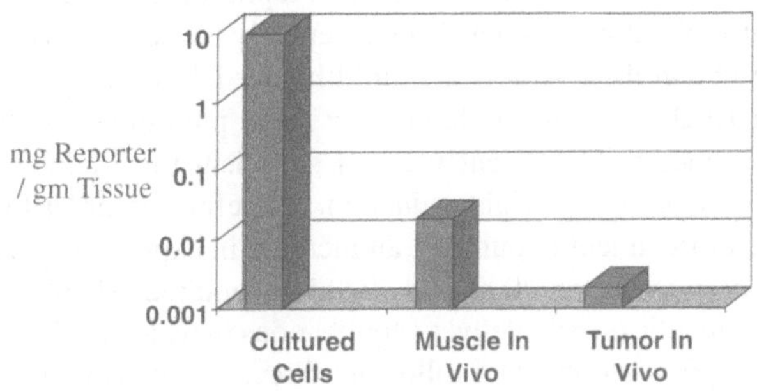

Fig. 3. In vivo efficacy of synthetic gene delivery systems is low

demonstrate pharmacological and physiological activities utilizing synthetic gene delivery systems. The following examples illustrate how increases in in vivo gene expression may expand the clinical gene therapy opportunities.

Erythropoietin is an extremely potent recombinant protein which is used to increase the endogenous production of red blood cells in patients that are experiencing anemia. In humans it is usually given once per week at 50–500 μg/dose. The corresponding dose of recombinant EPO in a mouse that leads to increased hematocrit is only 25–250 ng. The currently available naked DNA technology allows more than 250 ng of transgene product to be expressed and secreted from the site of a single intramusclular injection of plasmid DNA in a mouse. Therefore, it would be predicted that intramuscular injection of a plasmid secreting EPO from the muscle into the blood stream would lead to a sufficient amount of EPO to be expressed in order to increase the mouse hematocrit. Indeed, Tripathy showed convincingly that this approach works in the mouse, thus demonstrating the potential for using the muscle as a platform for secretion of a therapeutic protein into the blood (Tripathy et al. 1996). However, in order for this approach to work in humans, which have an approximately 2000-fold larger blood volume than mice, the expression level per administration needs to be increased by a corresponding 2000-fold. It remains to be determined whether this increased level of expression can be obtained in humans by simply increasing the DNA dose or whether more sophisticated improvements in the delivery system will be needed.

The results of in vivo experiments using leptin have led to similar conclusions. Intramuscular injection of a plasmid encoding leptin resulted in a sufficient amount of secreted leptin to stabilize weight gain and reduce blood glucose levels in genetically obese mice. However, in order to obtain these results a tenfold higher DNA dose was required than for EPO. Even at this higher dose, the physiological effect was marginal since the treatment was not sufficient to restore fertility or produce a significant weight reduction. Therefore, in order to transfer this type of treatment to humans, an increase in expression per administration of greater than 2000-fold would be required. There are several other therapeutic protein delivery applications in which the level of gene product expression obtained following direct in vivo administration of plasmid is either insufficient or just barely high enough to achieve a physiological response in mice. These include: insulin to treat type I diabetes and factor IX to treat hemophilia. In these cases it is even clearer that methods for increasing the level of gene product expression with synthetic gene delivery systems are required before practical dosage forms can be successfully evaluated in humans.

3.8 Overcoming the Barriers to More Efficient Gene Delivery

Thus, during the last 8 years we have witnessed rapid progress in the application of synthetic gene delivery systems; however, the applications remain limited today due to low efficiency of expression. The obstacles to efficient in vivo gene delivery and expression can be described in terms of "barriers" (Table 10). There are extracellular barriers and intracellular barriers. The extracellular barriers refer to those obstacles that the injected gene encounters before it reaches its target cell. The four main extracellular barriers are opsinins, phagocytes, extracellular matrix and degradative enzymes. Opsinins are proteins that attach themselves to the gene or the delivery system, inactivating the gene and its carrier. Phagocytes are cells that can seek out, engulf and actively digest the delivery system. The extracellular matrix is a zone of polymerized protein and carbohydrate that is present between cells protecting the plasma membrane of the target cell and it can be difficult for a relatively large DNA carrier system to pass through this barrier. And finally, DNAses are present in the serum and extracellular fluid that can rapidly digest unprotected DNA.

The principle intracellular barriers are the plasma membrane, the endosome and the nuclear membrane. Once the gene delivery system reaches its target cell it encounters the plasma membrane, which must be traversed before the gene can be expressed. Under the right circumstances, after the delivery system attaches to the plasma membrane it

Table 10. Improving gene delivery and expression: overcoming barriers to efficient gene delivery

Extracellular barriers	Opsinins; phagocytic cells; extracellular matrix; digestive enzymes
Intracellular barriers	Plasma membrane; endosome/lysosomes; nuclear membrane

can be taken into the cell by a process called endocytosis, but then the delivery system must have a mechanism for escaping from the endosome to avoid being degraded in the lysosomal compartment. Finally, if the gene is able to cross these barriers and enter the cell cytoplasm, it must still have a means of getting across the nuclear membrane.

Delivery system technology development directed at overcoming these barriers involves aspects of molecular biology, DNA condensation technology, and ligand conjugation chemistries. With respect to molecular biology, investigators are looking for ways to increase the level of gene product expressed from plasmids by optimizing promoters, enhancers, introns, terminator sequences and codon usage. Some investigators are also exploring ways to allow replication of the plasmid inside the nucleus of transfected cells. This approach would theoretically increase the number of plasmid copies per cell and also allow expression to persist at high levels within populations of rapidly dividing cells.

DNA condensation is another active area of scientific research. Today the principle ways of accomplishing condensation and packaging of plasmid DNA is with either hydrophobic cations or with hydrophilic polycations. Hydrophobic cations form liposomes or micelles which can interact with DNA and reorganize into a cationic lipid-DNA complex called "lipoplex." The hydrophilic polycations also form complexes when mixed with DNA and these complexes are called "polyplex." The hydrophilic polymers are of two general types, linear polymers such as polylysine and spermine, and the branched chain, spherical or globular polycations such as polyethyleneimine or dendrimers. An active area of scientific research involves understanding and controlling the DNA condensation and packaging processes with these agents, and determining the structure of the complexes.

Following packaging of the DNA by these methods investigators are interested in improving gene delivery efficiency by incorporating ligands into the complexes. These ligands are intended to introduce biological functions into the complexes in order to make them more effective at delivering genes into the target cells. Such ligands may include:

1. Peptides which have a specific cell surface receptor so that the complexes will be targeted to specific cells bearing this receptor

2. Nuclear localization signals so that DNA can enter the nucleus more efficiently
3. pH-sensitive ligands to encourage more efficient endosomal escape
4. Steric stabilizing agents to avoid interaction with biological factors which would destabilize the complexes after introduction into the biological milieu

Better methods for studying the biodistribution of plasmid in vitro and in vivo will facilitate synthetic gene delivery system research. A probe which would allow the simultaneous localization of plasmid DNA and gene product expression at the cellular and subcellular level would offer a means to better understand the cellular and molecular barriers to DNA delivery and should provide new insights leading to more effective plasmid delivery systems. To date, no technology has been reported which can enable the biodistribution of functionally and conformationally intact plasmid to be followed in real time in living cells, while simultaneously monitoring gene expression from the same plasmid. For this purpose, we recently described a nonperturbing plasmid labeling procedure to generate a highly fluorescent plasmid DNA preparation.

In order to create this system, we used the property of peptide nucleic acids (PNA) to hybridize to nucleic acids in a high-affinity and sequence-specific manner. A fluorescent PNA conjugate was hybridized to its complimentary sequence on a plasmid. The PNA binding site was cloned into a region of the plasmid that was not essential for transcription so that PNA binding would not interfere with expression. Fluorescent plasmid prepared in this way was neither functionally nor conformationally altered. PNA binding was sequence-specific, saturable, and extremely stable and did not influence the nucleic acid intracellular distribution. This method was utilized to study conformationally intact plasmid DNA biodistribution in living cells after cationic lipid-mediated transfection. A fluorescent plasmid expressing green fluorescent protein (GFP) enabled simultaneous localization of both plasmid and expressed protein in living cells and in real-time. GFP was expressed only in cells containing detectable nuclear fluorescent plasmid. This detection method offers a way to simultaneously monitor the intracellular localization and expression of plasmid DNA in living cells. This tool should aid in elucidation of the mechanism of plasmid delivery and its

Table 11. In vivo plasmid biodistribution: typical lipofection

1 000 000 plasmids/cell transfected
300 000 plasmids/cell in pellet
50 000 plasmids/cell intranuclear
1 000 plasmids/cell intranuclear

nuclear import with synthetic gene delivery systems, and in developing more efficient synthetic gene delivery systems.

Using this system we were also able to quantitate uptake of plasmid transfected into cultured cells. Southern blot analysis showed that, in a typical transfection, cells took up 30% of the input plasmid; 80%–90% of the plasmid that was isolated in the cell pellet could be removed from the cell surface with a dextran sulfate wash; 5% of the total input plasmid was taken up into the cells and was stable for at least 48 h in the cell, and about 30% of the intracellular plasmid was intranuclear after 48 h (Table 11).

In addition to providing a means to obtain biologically active fluorescent DNA, this approach utilizing PNA provides a way to introduce new physical and biological elements onto DNA without perturbing its transcriptional activity. This "gene chemistry" approach will be used in the future to couple ligands (e.g., nuclear localization signals or other peptides) onto the DNA in order to improve its in vivo bioavailability and expression.

3.9 Summary

It has been 40 years since the first studies showing functional delivery of purified nucleic acid into cells. These first studies showed that an intact virus particle was not necessary to produce live, infectious virus, but that only the viral nucleic acid was necessary. The synthetic gene delivery methods developed in those early years were important to the success of the recombinant protein science and industry, which still use these methods as routine laboratory tools. The retrovirus vectors which were the first delivery vehicles for gene therapy were generated by using these nonviral gene delivery techniques to introduce the desired genes into the retroviral packaging cell. And finally, a new generation of

Table 12. Improving gene delivery and expression: what is next?

Physical and structural aspects	Characterization and control
Characterize biological barriers	In vitro and in vivo mechanisms
Chemically modify carrier	New condensing agents; ligand conjugates
Chemically modify DNA	Medicinal chemistry of genes
Molecular biology aspects	Increase level and duration of expression; on/off control (gene switch)

synthetic gene delivery systems is being used today to directly administer genes into patients for a host of gene therapy applications.

There has been impressive progress in the development and application of synthetic gene delivery systems, but there is a long way to go before we will get control over the many variables that influence the activity and practical applicability of these systems (Table 12). A better scientific understanding in five major areas will lead to greatly improved synthetic gene delivery systems in the future. First, the process of the self-assembly of these synthetic systems needs to be better understood, so that the production of these complexes can be more precisely characterized and reproducibly controlled. Second, the in vivo obstacles to more efficient gene delivery need to be better understood so that these biological barriers can be rationally attacked. Third, new chemistries and delivery systems need to be explored in order to identify better ways to overcome these biological barriers. Fourth, approaches to chemically modify the DNA in ways that do not decrease transcriptional activity need to be explored in order to make delivery of the DNA more efficient. And finally, modifications in the DNA gene sequence need to be further developed in order to further increase the level of gene product expressed per cell, and to increase and control the level of gene product expressed.

References

Alexander HE, Koch G, Moran-Mountain I, Sprunt K, Van Damme O (1958) Infectivity of ribonucleic acid of poliovirus on HeLa cell monolayers. J Exp Med 108:493–506

Avery OT, MacLeod CM, MacCarty M (1944) Studies on the chemical nature of the substance inducing transformation of pneumococcal types. J Exp Med 79:137–158

Benvenisty N, Reshef L (1986) Direct introduction of genes into rats and expression of the genes. Proc Natl Acad Sci USA 83:9551–9555

Bouchard L, Gelinas C, Asselin C, Bastin M (1984) Tumorigenic activity of polyoma virus and SV40 DNAs in newborn rodents. Virology 135:53–64

Boussif O, Lezoualch F, Zanta MA, Mergny MD, Scherman D, Demeneix B, Behr JP (1995) A versatile vector for gene and oligonucleotide transfer into cells in culture and in vivo: polyethylenimine. Proc Natl Acad Sci USA 92:7297–7301

Brigham KL, Meyrick B, Christman B, Magnuson M, King G, Berry LC Jr (1989) In vivo transfection of murine lungs with a functioning prokaryotic gene using a liposome vehicle. Am J Med Sci 298:278–281

Dubensky TW, Cambell BA, Villarreal LP (1984) Direct transfection of viral plasmid DNA into the liver or spleen of mice. Proc Natl Acad Sci USA 81:7529–7533

Felgner PL, Rhodes G (1991) Gene therapeutics. Nature 349:351–352

Felgner PL, Ringold GM (1989) Cationic liposome-mediated transfection. Nature 337:387–388

Felgner PL, Gadek TR, Holm M, Roman R, Chan HW, Wenz M, Northrop JP, Ringold GM, Danielsen M (1987) Lipofection: a highly efficient, lipid mediated DNA-transfection procedure. Proc Natl Acad Sci USA 84:7413–7417

Felgner PL, Barenholz Y, Behr JP, Cheng SH, Cullis P, Huang L, Jessee JA, Seymour L, Szoka F, Thierry AR, Wagner E, Wu G (1997) Nomenclature for synthetic gene delivery systems. Hum Gene Ther 8:511–512

Felgner PL, Zaugg RH, Norman JA (1995) Synthetic recombinant DNA delivery for cancer therapeutics. Cancer Gene Ther 2:61–65

Fraley R, Subramani S, Berg P, Papahadjopoulos D (1980) Introduction of liposome-encapsulated SV40 DNA into cells. J Biol Chem 255:10431–10435

Friedmann T, Roblin R (1972) Gene therapy for human genetic disease? Science 175:949–955

Fung YK, Crittenden LB, Fadly AM, Kung HJ (1983) Tumor induction by direct injection of cloned v-src DNA into chickens. Proc Natl Acad Sci USA 80:353–357

Gould-Fogerite S, Mazurkiewicz JE, Raska K Jr, Voelkerding K, Lehman JM, Mannino RJ (1989) Chimerasome-mediated gene transfer in vitro and in vivo. Gene 84:429–438

Graham FL, Van Der Eb AJ (1973) A new technique for the assay of infectivity of human adenovirus 5 DNA. Virology 52:456–467

Hakim I, Levy S, Levy R (1986) A nine-amino acid peptide from IL-1beta augments antitumor immune responses induced by protein and DNA vaccines. J Immunol 157:5503–5511

Holland JJ, McLaren LC, Syverton JT (1959) The Mammalian cell-virus relationship III Poliovirus production by non-primate cells exposed to poliovirus ribonucleic acid. Proc Soc Exp Biol Med 100:843–845
Holt CE, Garlick N, Cornel E (1990) Lipofection of cDNAs in the embryonic vertebrate central nervous system. Neuron 4:203–214
Hwang LS, Gilboa E (1984) Expression of genes introduced into cells by retroviral infection is more efficient than that of genes introduced into cells by DNA transfection. J Virol 50:417–424
Israel MA, Chan HW, Martin MA, Rowe WP (1979) Molecular cloning of polyoma virus DNA in Escherichia coli: oncogenicity testing in hamsters. Science 205:1140–1142
Johnston SA, Anziano PQ, Shark K, Sanford JS, Butow RA (1988) Mitochondrial transformation in yeast by bombardment with microprojectiles. Science 240:1538–1541
Kaneda Y, Iwai K, Uchida T (1989) Increased expression of DNA cointroduced with nuclear protein in adult rat liver. Science 243:375–378
Mann R, Mulligan RC, Baltimore D (1983) Construction of a retrovirus packaging mutant and its use to produce helper-free defective retrovirus. Cell 33:153–159
Minson AC, Wildy P, Buchan A, Darby G (1978) Introduction of the herpes simplex virus thymidine kinase gene into mouse cells using virus DNA or transformed cell DNA. Cell 13:581–587
Mulligan RC, Howard BH, Berg P (1979) Synthesis of rabbit beta-globin in cultured monkey kidney cells following infection with a SV40 beta-globin recombinant genome. Nature 277:108–114
Nabel EG, Plautz G, Boyce FM, Stanley JC, Nabel GJ (1989) Recombinant gene expression in vivo within endothelial cells of the arterial wall. Science 244:1342–1344
Nabel GJ, Felgner PL (1993) Direct gene transfer for immunotherapy and immunization. Trends Biotechnol 11:211–215
Nabel GJ, Yang ZY, Nabel EG, Bishop K, Marquet M, Felgner PL, Gordon D, Chang AE (1995) Direct gene transfer for treatment of human cancer. Ann NY Acad Sci 772:227–231
Nicolau C, Le Pape A, Soriano P, Fargette F, Juhel MF (1983) In vivo expression of rat insulin after intravenous administration of the liposome-entrapped gene for rat insulin I. Proc Natl Acad Sci USA 80:1068–1072
Seeger C, Ganem D, Varmus HE (1984) The cloned genome of ground squirrel hepatitis virus is infectious in the animal. Proc Natl Acad Sci USA 81:5849–5852
Smull CE, Ludwig EH (1962) Enhancement of the plaque-forming capacity of poliovirus ribonucleic acid with basic proteins. J Bacteriol 84:1035–1040

Southern PJ, Berg P (1982) Transformation of mammalian cells to antibiotic resistance with a bacterial gene under control of the SV40 early region promoter. J Mol Appl Gen 1:327–341

Tang MX, Redemann CT, Szoka FC Jr (1996) In vitro gene delivery by degraded polyamidoamine dendrimers. Bioconjug Chem 7:703–714

Tripathy SK, Svensson EC, Black HB, Goldwasser E, Margalith M, Hobart PM, Leiden JM (1996) Long-term expression of erythropoietin in the systemic circulation of mice after intramuscular injection of a plasmid DNA vector. Proc Natl Acad Sci USA 93:10876–10880

Ulmer JB, Donnelly JJ, Parker SE, Rhodes GH, Felgner PL, Dwarki VJ, Gromkowski SH, Deck RR, DeWitt CM, Friedman A et al (1993) Heterologous protection against influenza by injection of DNA encoding a viral protein. Science 259:1745–1749

Vaheri A, Pagano JS (1965) Infectious poliovirus RNA: a sensitive method of assay. Virology 27:434–436

Wagner E, Zenke M, Cotten M, Beug H, Birnstiel ML (1990) Transferrin-polycation conjugates as carriers for DNA uptake into cells. Proc Natl Acad Sci USA 87:3410–3414

Wang CY, Huang L (1987) pH-sensitive immunoliposomes mediate target-cell-specific delivery and controlled expression of a foreign gene in mouse. Proc Natl Acad Sci USA 84:7851–7855

Wolff JA, Malone RW, Williams P, Chong W, Acsadi G, Jani G, Felgner PL (1990) Direct gene transfer into mouse muscle in vivo. Science 247:1465–1468

Wu CH, Wilson JM, Wu GY (1989) Targeting genes: delivery and persistent expression of a foreign gene driven b mammalian regulatory elements in vivo. J Biol Chem 264:16985–16987

Wu GY, Wu CH (1987) Receptor mediated in vitro gene transformation by a soluble DNA carrier system. J Biol Chem 262:4429–4432

Yang NS, Burkholder J, Roberts B, Martinell B, McCabe D (1990) In vivo and in vitro gene transfer to mammalian somatic cells by particle bombardment. Proc Natl Acad Sci USA 87:9568–9572

4 Tumor Suppressor Genes

B.E. Weissman

4.1 Introduction

The isolation of the first human tumor suppressor gene in 1986 fueled an immediate interest in gene replacement therapy as a novel treatment modality for human cancers (Friend et al. 1986) . The functional groundwork for the efficacy of this avenue of approach came from studies on the genetics of cancer using somatic cell genetics. The first report, in 1969, of the suppression of malignancy in hybrid cells between tumorigenic and nontumorigenic mouse cells provided evidence that normal cells possess genetic information capable of reversing many transformed features of tumor cells (Harris et al. 1969) . Since that initial study, many investigators have shown that introduction of normal genetic information into human cancer cells can cause suppression of cell growth in vitro and in vivo (Stanbridge 1992). Thus, the challenge facing scientists interested in the development of cancer gene therapies lies in the optimal delivery of potent tumor suppressor genes into tumor cells in vivo which can render them quiescent or prime them for destruction by other methods. This chapter will cover the identification of known tumor suppressor genes as well as the strategies to isolate novel tumor suppressor genes with different mechanisms of action.

Many investigations into human cancer during the past several years have focused on the molecular basis of its development. Although these studies have documented the many facets of cellular transformation, the event(s) responsible for many of these changes remain ill-defined. Recent reports have delineated some important changes in this process including alterations in proto-oncogene expression and differentiation patterns. However, the loss or inactivation of specific genetic information during human neoplastic progression has emerged as one of the most interesting yet poorly understood paradigms. Indeed, without an understanding of these molecular events, the relative contributions of genetic susceptibility and environmental factors to the development of human cancer remain nebulous.

While the number of known human oncogenes increased substantially during the past decade, the identification of tumor suppressor genes remains an arduous process. Despite the intense efforts of many laboratories, molecular studies have isolated only a limited number of putative tumor suppressor genes. The normal functions of these tumor suppressor genes and their effects upon introduction into tumor cells cover a broad range of cell activities. Therefore, to avoid confusion in this chapter, I will define the term "tumor suppressor" gene as any gene lost or inactivated during the development of human cancer.

4.1.1 Characterization of Tumor Suppressor Genes by Somatic Cell Genetics

The earliest studies suggesting the existence of functional tumor suppressor genes came from somatic cell genetics. The body of evidence, including our own studies, have shown that hybrids between tumor cells and their normal counterparts failed to form tumors upon inoculation into appropriate animals (Weissman 1990). These results were consistent with the presence of tumor suppressor information in the normal parent cell line. Indeed, we and others have demonstrated that loss of specific chromosomes from the normal parent correlated with the reexpression of tumorigenicity in the hybrid cells (Benedict et al. 1984; Stanbridge et al. 1981). Somatic cell hybrid studies have also specified a second class of tumor suppressor genes which control cellular immortality (Pereira-Smith and Smith 1983) . Thus, the majority of hybrids between human tumor cells and normal fibroblasts will proliferate for a limited number of population doublings in culture (Pereira-Smith and Smith 1983) . A more recent report also demonstrated a loss of the ability for gene amplification in human cell hybrids, segregating independently of these other two phenotypes (Tlsty et al. 1992).

One question raised by these studies asked whether a single or multiple tumor suppressor genes participate in the control of tumorigenicity. To address this possibility, we tested the tumorigenic potential of hybrids between different human tumor cells. Several hybrids of these tumor pairs were suppressed for tumorigenicity indicating the presence of two different tumor suppressor genes (Pasquale et al. 1988; Weissman and Stanbridge 1983). In addition, other complementation analyses indicate that at least four different groups for cellular senescence and three separate groups for tumorigenicity exist in human cells (Geiser et al. 1989; Pasquale et al. 1988; Weissman and Stanbridge 1983). Do any of these complementation groups correspond with known tumor suppressor genes? Expression of several genes including DCC, NF-1 and WT1 failed to correlate with the ability to suppress tumorigenicity in our somatic cell hybridization studies. Cell lines with inactivated p53 genes may form nontumorigenic hybrids while attempts to correlate RB expression with tumor suppression remains uninformative (Chen et al. 1994). Therefore, we cannot specify a known tumor suppressor gene as the defect associated with a complementation group.

While whole cell hybrids have provided initial data about the existence and location of tumor suppressor genes, they lack the sensitivity required for mapping studies. Precise chromosome mapping analyses require a method for the transfer of smaller amounts of genetic material from the normal parental cell line. The development of the microcell hybridization procedure fills this need by its ability to introduce a single chromosome into a recipient cell line (Ege and Ringertz 1974; Fournier and Ruddle 1977). Thus, suppression of tumorigenicity in a tumor cell line after introduction of a single human chromosome provides functional evidence for the location of a tumor suppressor gene. Therefore, this technique augments the previous cytogenetic studies as well as furnishing a novel method for the mapping of tumor suppressor genes.

Microcell hybridization studies have now identified tumor suppressor genes on at least nine different human chromosomes (Weissman and Conway 1995). In several cases, evidence from different studies suggests that at least two different tumor suppressor genes map to the same chromosome. Human chromosome 11 appears to possess at least three different tumor suppressor activities while chromosome 17 may contain at least four different tumor suppressors (Call et al. 1990; Dong et al. 1995; Dowdy et al. 1991; Gessler et al. 1990; Miki et al. 1994; Misra and Srivatsan 1989; Nigro et al. 1989; Steeg et al. 1988; Viskochil et al. 1990; Wallace et al. 1990). Whether these genes function independently of each other or participate in a common pathway remains an intriguing area for further investigation.

4.1.2 Isolation of Tumor Suppressor Genes

Several different trails have led to the identification of tumor suppressor genes. The RB gene was initially isolated because of its association with the development of the pediatric eye tumor retinoblastoma (Friend et al. 1986; Fung et al. 1987; Lee et al. 1987). The p53 gene was first identified due to its altered expression in human tumors (Crawford

1983). Originally classified as an oncogene, subsequent studies have demonstrated that the wild-type protein acts as a tumor suppressor gene (Eliyahu et al. 1989; Finlay et al. 1989). These genes show expression in virtually all normal tissues and participate in the development of a wide range of tumors from different developmental origins (Nigro et al. 1989; Weinberg 1989). Both these proteins are bound specifically by the transforming proteins of several DNA tumor viruses leading to their inactivation (Helin and Harlow 1993; Vogelstein and Kinzler 1992). Reports have linked these proteins to signal transduction, regulation of transcription, cell cycle control and genomic instability (Hartwell 1992; Helin and Harlow 1993; Pietenpol et al. 1990; Rotter et al. 1993; Vogelstein and Kinzler 1992). Isolation of other tumor suppressor genes have relied on loss of heterozygosity (LOH) studies and positional cloning including BRCA1, BRCA2 and APC (Groden et al. 1991; Kinzler et al. 1991a; Miki et al. 1994; Wooster et al. 1995). Functional studies have contributed significantly to the identification of several tumor suppressor genes such as PTEN and KAI-1 (Dong et al. 1995; Li et al. 1997; Steck et al. 1997). Finally, conservation of function between human and other species have aided in the isolation of the sMAD4 and Patched tumor suppressor genes (Hahn et al. 1996a,b; Johnson et al. 1996).

4.1.3 Functional Aspects of Tumor Suppressor Genes

Since the reports of the isolation of the RB gene in 1986, a small number of other tumor suppressor genes have emerged each year (Table 1). I have grouped them into somewhat arbitrary categories for the purpose of discussion of common functions. Clearly, many tumor suppressor genes normally contribute to many features of cellular growth, development and differentiation. As mentioned above, both the p53 and RB proteins play major roles in the control of cell division under normal conditions and in response to environmental damage. Other tumor suppressor genes fall into the family of cyclin-dependent kinase inhibitors (CDKIs) which regulate the activity of the RB protein (Xiong et al. 1992). The ATM gene, whose loss underlies the cancer-prone disease ataxia telangeictasia, appears to help regulate the cell's response to DNA damage by a variety of agents (Savitsky et al. 1995).

Table 1. Known tumor suppressor genes

Cell cycle control genes (ubiquitous expression)	RB, p53, $p16^{INK4a}$, $p15^{INK4b}$, $p19^{arf1}$, ATM
Cellular structure genes (differentiation-specific expression)	APC, NF-2, DCC
Transcriptional regulation (differentiation-specific expression)	WT-1, VHL, BRCA1(?), BRCA2(?)
Signal transduction (variable expression)	BRCA1, BRCA2(?), sMAD4, FHIT, KAI1, NF-1(?), NM-23 (?), PTEN/MMAC1
DNA repair (ubiquitous expression)	MSH2, MLH1, PTC

Other tumor suppressor genes appear to function as transcriptional regulators. The WT-1 protein, isolated by its association with Wilms' tumor, contains a zinc-finger motif which can control transcription of a number of growth factors (Drummond et al. 1992; Wang et al. 1992). The VHL tumor suppressor gene functions as a regulator of translation of several important genes including vascular endothelial growth factor (VEGF; Latif et al. 1993). Loss of this gene leads to the development of both sporadic and familial renal cell carcinomas (Latif et al. 1993). The tumor suppressor genes, BRCA1 and BRCA2, play important roles in familial ovarian and breast cancer development (Futreal et al. 1994; Miki et al. 1994). They appear to code for ring-finger proteins with potential transcriptional regulatory activities. Several recent reports have indicated that they physically associate with proteins involved in DNA repairs such as RAD52 (Zhang et al. 1998).

Another group of tumor suppressor genes may operate through signal transduction associated with normal development. The DCC gene was isolated because of its location on chromosome 18 which showed LOH in human colorectal carcinomas (Fearon et al. 1990) . It shares sequence homology with the N-CAM protein suggesting an involvement in cell adhesion (Fearon et al. 1990). It now appears that this gene participates in normal neural development. Loss of the APC gene contributes to the onset of both sporadic colorectal cancers and in tumors found in family with adematous polyposis coli (Groden et al. 1991; Kinzler et al. 1991a,b). This gene product forms part of the intercellular junctions found in most epithelial cells. Originally thought to represent a structural element, recent studies have indicated that these proteins also participate in cellular development. Another member of this group, NF-2, bears striking homology to the cytoskeletal proteins moesin, ezrin and radixin (Trofatter et al. 1993). The normal function of this gene may lie in the control of signal transduction pathways by physical manipulation of cell surface molecules such as growth factor receptors.

Some tumor suppressor genes represent members of known signal transduction pathways. The sMAD4 gene product helps transmit the growth inhibitory signal from the TGFβ-1 receptor to the nucleus after ligand binding (Hahn et al. 1996b). The PTEN/MMAC1 gene appears to possess phosphatase activity presumably exerting an anti-proliferative effect on cells by deactivation of signal transduction molecules (Li et al. 1997; Steck et al. 1997). Another tumor suppressor gene, NF-1, contributes to regulation of ras oncogene functions, another central proliferative and developmental signal transduction pathway (Viskochil et al. 1990; Wallace et al. 1990). Finally, a variety of human tumors display mutations in a family of genes involved in DNA mismatch repair (Kolodner 1996). While these genes fall into the broad classification of tumor suppressors, their utility as targets for gene therapy seem doubtful. They probably contribute to the generation of genomic instability associated with human tumors rather than specific defects in growth or differentiation control mechanisms.

As we have determined the pathways by which these genes regulate the normal growth and development of cells, we have recognized that the initiation of human cancer may require the loss of function of a pathway rather than a particular tumor suppressor gene. Recent reports suggest that loss of RB function may occur through three different mechanisms- inactivation of the RB gene itself, overexpression of the CDK4 gene or loss of function of the $p16^{INK4a}$ gene (Sherr 1996). One of these events occurs in almost 100% of human cancers indicating the critical importance of the loss of this cell cycle regulatory pathway in the progression of human cancer. Similar pathways may also exist for the p53 tumor suppressor gene.

In this brief introduction, it seems evident that multiple targets exist for gene replacement strategies for the treatment of human malignancies. As present, most studies have concentrated on the introduction of the RB and p53 genes into human tumors, as they exert a powerful growth inhibitory effect on these cells. In addition, the reexpression of these genes in tumor cells may render them more sensitive to the effects of adjuvant radiation therapy and chemotherapy. However, the requirement for introducing these genes into virtually 100% of the tumor cells in vivo in order to significantly block tumor growth remains a daunting obstacle. In the next sections of the chapter, I will relate two types of genetic studies aimed at the isolation of other types of tumor suppressor genes. The identification of these genes may provide better targets for gene replacement strategies because their suppressive effects may not depend on their introduction into 100% of the tumor cells.

4.2 The Search for a Novel Tumor Suppressor Gene for Human Epithelial Malignancies

The identification of the chromosomal locations of tumor suppressor genes has relied for the most part on molecular analyses using restriction fragment length polymorphisms (RFLPs) markers and cytogenetics. Indeed, isolation of the RB, DCC, MCC, APC, NF-1, NF-2 and WT-1 genes by positional cloning directly depended on the knowledge of their chromosomal sites. Many studies have suggested that LOH will predict the location of all deleted or inactivated tumor suppressor genes. However, our data indicate that LOH studies may miss the location of other tumor suppressor genes, a conclusion supported by other investigator's studies (Bader et al. 1991; Loh et al. 1992; Yamada et al. 1990). For example, Oshimura and co-workers have demonstrated tumor suppressor activity on chromosomes 6 and 9 for endometrial carcinoma, sites unanticipated by molecular analyses (Yamada et al. 1990). Bader et al. (1991) have mapped non-p53 tumor suppressor information for neuroblastomas to chromosome 17. In some cases, the use of one or two RFLP markers missed LOH when restricted to a particular area of the chromosome (Koufos et al. 1989; Reeve et al. 1989). In addition, the minimum significance becomes obscure when all chromosomes show some degree of LOH (Vogelstein et al. 1989). Finally, lack of functional evidence for the presence of a tumor suppressor gene remains the major drawback to the isolation of novel tumor suppressor genes.

4.2.1 Tumor Suppressor Genes in Squamous Cell Carcinomas

One notable finding from our complementation studies showed that adult and pediatric cancers fell into different groups (Pasquale et al. 1988). We have devoted most of our characterization and isolation efforts to the latter group of tumors. Recently, we have initiated studies on the genetics of squamous cell carcinomas. Many reports have established the importance of the loss of activity of genes such as RB, p53 and p16^{INK4a} genes in the development of these tumors (Huang et al. 1993; Kratzke et al. 1996; Muktar and Bickers 1993; Reed et al. 1996). However, these changes only occur in approximately 25% of these cancers, implying the existence of other tumor suppressor genes. In our initial studies, we have tested the hypothesis that human chromosome 11 carries a tumor suppressor gene for human squamous cell carcinomas (Conway et al. 1992). Our results clearly established the presence of a functional tumor suppressor gene on this chromosome. RFLP analyses of squamous cell carcinoma tumor material indicated a loss of heterozygosity on chromosome 11 for this malignancy (Heo et al. 1989). Whether this putative gene implicated by molecular studies corresponds to the functional tumor suppressor gene of our studies remains a central focus of our research program.

4.2.2 A Common Tumor Suppressor Gene for Many Human Carcinomas?

Several lines of evidence strongly implicate the existence of a tumor suppressor gene on the long arm of chromosome 11 which undergoes inactivation in a large number of adult epithelial malignancies besides squamous cell carcinoma. The earliest example of a functional human tumor suppressor gene came from the studies of Stanbridge and coworkers, who showed that a functional tumor suppressor gene exists on chromosome 11 for the human cervical carcinoma cell line HeLa (Saxon et al. 1986). Monochromosome transfer experiments have additionally demonstrated tumor suppressor activity on this chromosome for cell lines derived from a different cervical cancer and two breast cancer cell lines (Koi et al. 1989; Negrini et al. 1992). Recent molecular studies have shown LOH for the long arm of chromosome 11 for breast cancer and cervical cancer (Hampton et al. 1994a,b). Thus, the isolation of a tumor suppressor gene for squamous cell carcinoma may clearly benefit the characterization of other types of human cancer.

4.2.3 Functional Analyses of Tumor Suppressor Activity in Squamous Cell Carcinomas

We have applied the technique of microcell hybridization to identify functional tumor suppressor genes for squamous cell carcinomas. We initially examined the effects of transfer of three different human chromosomes, 11, 13 and 15, on the in vivo growth potential of a human squamous cell carcinoma line, A388. This cell line forms large well-differentiated tumors upon inoculation into athymic nude mice (Giard et al. 1974). Chromosome 11 appears to carry a suppressor gene for several human adult epithelial malignancies while chromosome 13 contains the RB tumor suppressor gene (Friend et al. 1986; Fung et al. 1987; Lee et al. 1987; Weissman 1990). We opted for chromosome 15 as a negative control because previous studies had not shown LOH on this chromosome. We first characterized the in vitro growth and differentiation properties of the A388 cell line and the microcell hybrids. All cell lines displayed virtually identical phenotypes including population doubling times, soft agar growth and cell cycle distribution. To date, we have found that the microcell hybrids behave no differently than the A388 parental cell under standard tissue culture conditions.

In contrast, in vivo studies showed that introduction of chromosome 11 caused suppression of tumorigenicity of the A388 cell line in nude mice (Conway et al. 1992). Microcell hybrids containing either chromosome 13 or 15 grew as well as the A388 cell line under these conditions (Conway et al. 1992). Most importantly, the site of inoculation governed the degree of tumor suppression. Only injection of the cells into the most physiologically relevant site, the skin graft assay, resulted in complete suppression of tumorigenicity. Inoculation of the microcell hybrids into the standard subcutaneous site in nude mice led to a slower growing tumor. Thus, two important lessons arise from these results. First, the tumor suppressor activity on chromosome 11 did not display any activity on cells grown in culture in contrast to the dramatic effects observed by the introduction of the p53 or RB genes. Second, the choice of inoculation site for testing efficacy of a tumor suppressor gene replacement strategy for human tumor cells may profoundly affect the results.

While the use of the skin graft assay has proved useful for initial studies, it suffers from several drawbacks including the difficulty of the grafting procedure, the requirement for large numbers of cells, the length of the assay (2–4 months), and the expense of the animals. To circumvent these problems, we searched for a suitable in vitro model for measuring tumor suppressor activity in our cell lines. Because complete tumor suppression only occurred in vivo when we tested the cells in their normal growth site, we looked for in vitro models that most closely mimicked normal epidermal differentiation.

The most physiological relevant in vitro model for normal epidermal differentiation lies in the raft model (Asselineau and Prunieras 1984; Fusenig et al. 1983). In this system, epidermal keratinocytes growing on collagen gels containing fibroblasts are "floated" on the medium to give an air-liquid interface resembling the situation in vivo. Under these conditions, normal human epidermal keratinocytes (NHEKs) form keratinizing squamous epithelium with clearly defined basal, spinosum, granulosum and corneum layers. Most biochemical and immunological markers of epidermal differentiation appear in the appropriate layers. The degree to which the raft assay resembles normal in vivo growth depends on the origin of the fibroblasts embedded in the collagen layer (Wilson et al. 1992). In our assay, we used normal human dermal fibroblasts to try to provide the most "physiological" model. These cells support the growth of the NHEKs with an apparently normal-looking pattern of epidermal differentiation (Gioeli et al. 1997).

We tested our parental and control cells as well as the nontumorigenic microcell hybrids for their phenotype in the raft assay. The parental A388 cell line grew into a multilayer (8–10 layers) structure with abnormal cell morphology and squamous differentiation. Transfection of the NPT gene for neomycin-resistance or the introduction of a normal human chromosome 13 did not alter this pattern. However, the nontumorigenic microcell hybrids containing a normal copy of chromosome 11 failed to proliferate in the raft assay (Gioeli et al. 1997). Thus, the growth of these cell lines in the raft assay correlates exactly with their tumorigenic potential in vivo.

4.2.4 The Squamous Cell Carcinoma Tumor Suppressor Gene Does Not Affect Cell Cycle Control

We next determined whether the nontumorigenic microcell hybrids showed any proliferative capacity in the raft assay. Based on previous studies on the actions of tumor suppressor genes, it seemed likely that the chromosome 11 microcell hybrids would show either growth arrest and/or increased apoptosis upon exposure to the raft assay. We followed the course of cell growth over a 28 day period, the maximum time the rafts remain healthy in culture. The parental A388 cell line formed multiple layers by 5 days and continued to expand until day 14 (Gioeli et al. 1997). By day 21, it appeared that the health of the cells had begun to decline, presumably due to the loss of viability of the fibroblasts. In contrast, the nontumorigenic microcell hybrids only produced two-layer growth by day 5 that became unhealthy by day 21. These results appeared consistent with a model in which the microcell hybrids underwent a growth arrest event after one population doubling on the rafts.

To determine if the reduced growth arose from a decrease in cell proliferation, we determined the proliferation index as measured by BrdU incorporation. A388 showed BrdU staining in both the basal and suprabasal layers, consistent with results found in other squamous cell carcinomas. We predicted that fewer BrdU-positive cells would appear in the chromosome 11 microcell hybrids, indicative of a reduced proliferation index. However, while the microcell hybrids showed a slight decrease in BrdU incorporation compared with A388 at the earliest time points, the observed proliferation indices appeared nearly equivalent during the remainder of the time course (Gioeli et al. 1997). Thus, it seemed like lack of growth of the microcell hybrid did not result from growth arrest.

Because no significant difference in proliferation indices occurred among the cell lines, we tested whether increased apoptosis could account for the reduced growth of the chromosome 11 microcell hybrids. Examination of hematoxylin and eosin (H and E) stained sections for apoptotic bodies suggested similar, if not greater amounts of cell death in A388 and controls than in chromosome 11 microcell hybrids (Gioeli et al. 1997). We confirmed this result using the TUNEL (TdT-mediated dUTP Nick-End Labelling) assay to label free DNA ends, a signature of apoptotic (and necrotic) cell death (Gioeli et al. 1997). Although the percent of apoptosis was higher in the microcell hybrids than in A388 cells at early time points, we observed nearly equivalent indices of apoptosis at the later time points (Gioeli et al. 1997). These data suggested that increased cell death did not account for the decreased cell growth on the organotypic raft cultures. Therefore, we postulate that the operative tumor suppressor gene on chromosome 11 for the squamous cell carcinoma cell line acted by a mechanism other than restoration of cell cycle control and/or apoptosis.

4.3 Identification of Metastasis Suppressor Genes

Most of the tumor suppressor genes identified to date presumably act at the early stages of tumor development. Introduction of these genes into tumor cells causes inhibition of growth in culture and/or in animal models. However, in most solid tumors, metastatic disease, rather than the primary tumor, eventually causes mortality. Several studies have shown that acquisition of metastatic potential results from accumulation of several molecular defects distinct from initial tumor formation (Bishop 1987; Bremmer and Balmain 1990). Thus, characterization of the later molecular events in aggressive tumor cells could present several potential benefits including an understanding of the interplay between different tumor suppressor genes and cell cycle genes like p53, the contributions of genetic susceptibility, and the impact of environmental factors on the development of metastasis. Therefore, knowledge of genetic loci whose loss or inactivation contributes to metastasis development could help with decisions of treatment and prognosis. Furthermore, replacement gene therapy aimed at the introduction of metastasis suppressor genes could present an early intervention to present the spread of tumor cells to distant sites while other treatment modalities attack the primary tumor.

4.3.1 Chromosome 11 Contains a Metastasis Suppressor Gene for Breast Cancer

Loss of heterozygosity studies have shown that regions on chromosome 11 frequently undergo loss during the development of breast tumors (Devilee et al. 1991; Hampton et al. 1994a; Negrini et al. 1994, 1995;

Winquist et al. 1993). These genetic alterations suggest the presence of tumor suppressor genes whose inactivation allows tumor initiation and metastases. To date, several tumor suppressor genes have been identified on chromosome 11. Both a tumor suppressor gene for Wilms' tumor (WT1) and a metastasis suppressor gene for prostate cancer (kai-1) map to chromosome 11p13 and 11p11, respectively (Call et al. 1990; Dong et al. 1995; Gessler et al. 1990). We and others are searching the 11p15.5 region for a second sporadic Wilms' tumor suppressor gene and perhaps an embryonal rhabdomyosarcoma tumor suppressor gene (Hao et al. 1993; Reid et al. 1996). Additional genes implicated in the development of malignancy include the β2-integrin gene and the ATM gene (Savitsky et al. 1995; Zutter et al. 1995).

Many studies have demonstrated that the introduction of normal chromosomes into malignant cell lines may restore gene function and reverse the transformed phenotype. As mentioned above, monochromosomal transfer of human chromosome 11 suppressed the tumorigenic phenotype of several tumor lines including the MCF-7 human breast cancer cell line (Bader et al. 1991; Conway et al. 1992; Weissman et al. 1987) . However, loss of separate genes on chromosome 11 might contribute to different stages of cancer. For example, cytogenetic analyses of metastatic breast tumors have shown that chromosome 11 rearrangements often appear as late events in breast cancer, suggesting the presence of a metastasis suppressor gene (Trent et al. 1993).

In order to determine if a gene on chromosome 11 might alter the tumorigenic or metastatic phenotype of a human breast carcinoma cell line, we introduced neo-tagged chromosomes into the MDA-MB-435 breast cancer cell line. The MDA-MB-435 cell line forms tumors when inoculated into the mammary fat pads of nude mice. The cells will also metastasize to the lungs of the animals, providing a useful model for spontaneous metastasis similar to the process in human disease. MDA-MB-435 cells and four chromosome 11 microcell hybrids formed tumors following injection, indicating that chromosome 11 did not encode a tumor-suppressor gene for this tumor type (Phillips et al. 1996). However, the incidence of lung metastases (macroscopic and microscopic) from the four neo11/435 microcell hybrids was significantly different from that of the metastatic parent. MDA-MB-435 cells metastasized to the lung in 93% of the mice, with a mean number of 20±6 (S.E.M.) metastases per animal. In contrast, all four of the chromosome

11 microcell hybrids displayed a reduced incidence of metastases in the lung and extrapulmonary sites, with a mean number of less than two lung metastases per animal (Phillips et al. 1996). These results indicated the presence of a metastasis suppressor gene for the MDA-MB-435 breast carcinoma cell line on chromosome 11.

4.3.2 Characterization of the Kai-1 Metastasis Suppressor Gene in Human Breast Cancer

Recent efforts have led to the isolation of a few metastasis suppressor genes for human cancers (Dear and Kefford 1990). The nm23 gene regulates metastatic potential without affecting tumorigenicity (Steeg et al. 1988). However, this gene maps to the long arm of chromosome 17. The Kai-1 gene, also known as CD82 (C33 antigen), maps to 11p11.2 and codes for a glycoprotein in the transmembrane 4 super family (TM4SF) of proteins (Dong et al. 1995; Fukudome et al. 1992; Imai et al. 1992). Because the identification of the Kai-1 gene relied on its ability to suppress prostate cancer metastatic ability in vivo, it makes it an attractive candidate for a breast cancer metastasis suppressor gene. In addition, the gene for another TM4SF member, Tapa-1 (CD81), also maps to chromosome 11 in band p15.5 (Virtaneva et al. 1994). Although no studies have implicated Tapa-1 in cancer suppression, it can induce cell-cell adhesion, a property often altered in metastatic cells (Takahashi et al. 1990; Wright and Tomlinson 1994; Zutter et al. 1995). The functions of these TM4SF surface proteins remain largely unknown. Some associate with cell surface receptors implying a role in signaling complexes, while others may participate in maintenance of cell integrity, proliferation, and adhesion (Bradbury et al. 1992; Nojima et al. 1993). Therefore, alterations in expression levels of the KAI-1 and/or TAPA-1 proteins could result in acquisition of invasive or metastatic ability.

We wanted to determine if either KAI-1 or TAPA-1 protein might control human breast metastatic ability. We initially examined whether expression of the Kai-1 gene correlated with the metastatic abilities of several human breast cancer cell lines. We measured expression of this gene by northern blot analysis in six cell lines which ranged from metastatic (MDA-MB-435) to poorly metastatic (MDA-MB-231) to nonmetastatic (MCF-7, T47D, ZR-75). Expression of the Kai-1 mRNA inversely correlated with the metastatic ability of the cell lines (Yang et al. 1997). In contrast, the expression of estrogen receptors, progesterone receptors and in vitro invasiveness did not correlate with metastatic potential. Thus, low KAI-1 expression appeared to predict high metastatic potential as observed in the rat prostate tumor model.

To determine whether KAI-1 protein levels correlated with the metastatic potentials of the MDA-MB-435 breast cancer cells and the four suppressed chromosome 11 microcell hybrids, we utilized western blot analysis. We could detect little or no protein in the parental MDA-MB-435 cell line and one single-cell derived subclone (Phillips et al. 1998). However, each metastasis-suppressed chromosome 11 microcell hybrid showed an increased level of KAI-1 protein (Phillips et al. 1998). One of the hybrids expressed amounts of KAI-1 protein similar to a metastasis-suppressed rat prostate tumor microcell hybrid line possessing human chromosome 11 (Dong et al. 1995). These data showed that chromosome 11 introduction into the metastatic breast cell line resulted in increased KAI-1 protein expression correlating with metastasis suppression (Phillips et al. 1998).

To determine whether the genetic information responsible for the increased KAI-1 expression mapped to chromosome 11, we also examined four MDA-MB-435 microcell hybrids containing a normal copy of human chromosome 6 which remained metastatic (Phillips et al. 1996). Western blot analysis comparing the amount of KAI-1 protein in the parental cell lines with chromosome 6 and chromosome 11 microcell hybrids and neomycin transfectant controls demonstrated that the chromosome 11 microcell hybrids showed the highest levels of the KAI-1 protein (Phillips et al. 1998). Three of four chromosome 6 microcell hybrids displayed KAI-1 levels not significantly increased over the parental cells and neo-transfectants. Interestingly, the one chromosome 6 microcell hybrid which showed some increase in KAI-1 expression also demonstrated some reduction in metastatic potential (Phillips et al. 1996). We also used the same western blots to measure the expression of TAPA-1, the other potential metastasis suppressor gene on chromosome 11. Our results indicated only a slight increase in the levels of this protein among the metastasis-suppressed cell lines. Therefore, these data strongly suggest that KAI-1, and not TAPA-1, expression may account for the loss of metastatic potential in the chromosome 11 microcell hybrids.

4.3.3 Transfection of the Kai-1 Gene into the MDA-MB-435 Cell Line

We next directly tested whether increased expression of KAI-1 protein could account for the metastasis suppression seen in the chromosome 11 microcell hybrids. We introduced Kai-1 cDNA into the MDA-MB-435 cell line by standard transfection protocols. After isolation of colonies of presumed clonal origin, we used PCR to verify the transfer of the KAI-1 gene into each clone (Phillips et al. 1998). We then confirmed the expression of the transfected gene by western blot. Each clone showed an increased level of expression as compared to the MDA-MB-435 parental cell line (Phillips et al. 1998). However, the protein in several of the clones displayed a wider size range (40–75 kDa) than in the microcell hybrids. This observation suggested a more extensive post-translational modification of the KAI-1 protein, presumably heavier glycosylation. These data showed that, unlike our observations with the chromosome 11 microcell hybrids, the amount of KAI-1 protein and the degree of post-translational glycosylation vary greatly among the transfectants. Thus, the KAI-1 expression from an introduced vector appears qualitatively distinct from that seen after introduction of a whole, normal chromosome 11.

We next tested whether Kai-1 transfectants showed reduced metastatic potential by inoculation into mammary fat pads of athymic nude mice and measurement of lung metastases. As with most solid tumors, inherent heterogeneity for metastatic potential exists within the MDA-MB-435 parent cell population. We therefore isolated several single cell clones and assayed them for metastatic potential. Seven different subclones were inoculated into at least eight nude mice. Even though the subclones had a wide range of numbers of lung metastases per mouse, the incidence of metastasis was about the same as that for the parental cell line (Phillips et al. 1998). The incidence of metastasis for the Kai-1 transfectant clones and the neo control transfectants did not appear significantly different from the parental cells (Phillips et al. 1998). Again, only the chromosome 11 microcell hybrids showed a significantly lower incidence of lung metastases (Phillips et al. 1998). This result suggests that a gene other than the Kai-1 gene on chromosome 11 may control metastasis potential in the chromosome 11/MDA-MB-435 model. However, because the range of metastases in the transfectant group did appear smaller than the parental groups, the data suggest that Kai-1 does indeed have some suppression function.

However, an alternative explanation exists for the lack of significant metastasis suppression among the Kai-1 transfectants. Loss of the cDNA or of expression of the KAI-1 protein could explain the seeming disparity between the western data and the lack of metastasis suppression in the Kai-1 transfectants. To test this possibility, we analyzed locally growing tumors for KAI-1 protein expression as compared to the cells grown in culture. Western blot analysis of inoculated cells and locally growing tumors for three Kai-1 transfectants and the MDA-MB-435 parental cells showed a decrease in the amount of KAI-1 protein in each transfectant primary tumor (Phillips et al. 1998). We could detect little, if any, protein in the primary tumor or the metastatic lesions of the MDA-MB-435 cells (Phillips et al. 1998). However, the inoculated cells showed the same low level of KAI-1 expression. We also analyzed one Kai-1 transfectant for protein levels in the lung metastases where we observed a further decrease. A comparable analysis of three inoculated chromosome 11 microcell hybrids with paired locally growing tumor proteins showed equal or increased levels of KAI-1 for all hybrids (Phillips et al. 1998). These data show that KAI-1 expression in the transfectants undergoes reduction in vivo during tumor progression to metastasis, perhaps owing to the heavy glycosylation. The metastasis-suppressed chromosome 11 microcell hybrids do not show a decrease in KAI-1 expression in vivo. Taken together, the western blot, transfection, and in vivo expression data suggest that Kai-1 may represent the operative metastasis suppressor gene on chromosome 11 for breast cancer.

4.4 Concluding Remarks

The potential for the successful development of gene replacement therapies using tumor suppressor genes depends on several factors. One objective lies in the enhancement of vectors to ensure delivery to the maximum number of tumor cells in vivo. A second goal aims at prolonging the expression of the introduced tumor suppressor gene to provide sufficient time for eradication of the tumor cells. Theoretically, if reexpression of a tumor suppressor gene in a tumor cell causes a growth arrest and/or apoptosis in that cell alone, then 100% of the cells must receive the gene therapy vector to completely destroy the tumor. As this goal may prove difficult to achieve, the generation of a bystander effect might allow effective cancer gene therapy with only a modest frequency of transfection. For example, if the destruction of a tumor cell by apoptosis triggers cell death in the surrounding tumor cells, then one might have to introduce the tumor suppressor gene into 10% of the cells. A recent report has shown this phenomenon, in which introduction of an adenovirus-associated virus containing the p53 gene into a human non-small cell lung carcinoma cell line produced a dramatic reduction in tumor size in vivo (Qazilbash et al. 1997). However, the investigators estimate that, at most, 10% of the cells received the virus encoding p53. The investigators suggest that a bystander effect took place albeit by an unknown mechanism.

Our studies with the tumor suppressor gene for adult epithelial malignancies on chromosome 11 suggest that it exerts it effects through a mechanism other than cell cycle arrest or apoptosis. We believe that this tumor suppressor gene normally functions in the process which controls differentiation. In the case of the squamous cell carcinomas, restoration of the tumor suppressor activity appears to restore the cells to a basal cell phenotype, a cell layer normally restricted to one to two layers. Therefore, the microcell hybrids still divide like normal basal cells but lack the additional signals necessary for undergoing full epidermal differentiation. Our preliminary data suggest that the introduction of the operative tumor suppressor gene leads to the reexpression of an autocrine inhibitory factor. Thus, treatment of the parental tumor cells with conditioned medium from the microcell hybrids causes an inhibition of growth in the raft assay similar to the microcell hybrid. If this observation holds true, then introduction of this tumor suppressor gene by gene therapy methods into a tumor in vivo would produce an inhibitory factor. Therefore, one cell receiving the tumor suppressor gene could lead inhibition of growth of the surrounding cells. We are currently testing this hypothesis by assessing the growth of mixtures of the A388 cell line and the chromosome 11 microcell hybrids in the raft assay.

The lessons from our studies with the Kai-1 gene indicate the caveats of strategies for effective introduction of tumor suppressor genes into tumor cells. In this case, it appears that suppression of metastasis can occur when this gene enters the cells under the control of its normal genetic components. However, metastasis suppression may only occur when KAI-1 protein levels rise above a threshold and/or when the protein maintains the proper level of post-translational modification. Our results indicate that transfection of the gene in an expression vector under an exogenous promoter may not fulfill these requirements. These observations impact upon the strategies for cancer gene therapy. The successful replacement of normal tumor suppressor gene function into human tumor cells by gene therapy will depend on their normal levels of expression and protein modifications. Therefore, introduction of the RB or BRCA1 gene into human cancer cells under the control of exogenous promoters could lead to inappropriate expression of the proteins. Furthermore, the phosphorylation status of these proteins is connected to their normal functions. If these parameters change by virtue of the transfection process, the tumor suppressive effects of these proteins may suffer.

In summary, the increasing availability of the human genomic sequence as well as the renewed interest in cancer genetics should lead to the identification of a large number of new tumor suppressor genes during the next decade. While the effort of many scientists in the area of cancer gene therapy will lead to better and more efficient delivery systems, the mechanisms of action of these new tumor suppressor genes could provide more efficacious targets for intervention in the treatment of cancer. In addition, advances in our understanding of the limitations of the current model systems for testing should direct us in a more rapid fashion to the development of these new modalities. Continued interactions between the research groups that develop the vectors and those that isolate novel tumor suppressor genes should ensure expeditious progress in the field of cancer gene therapy.

References

Asselineau D, Prunieras M (1984) Reconstruction of simplified control of fabrication. Br J Dematol [Suppl] 111:219–211

Bader SA, Fasching C, Brodeur GM et al (1991) Dissociation of suppression of tumorigenicity and differentiation in vitro effected by transfer of single human chromosomes into human neuroblastoma cells. Cell Growth Differ 2:245–255

Benedict WF, Weissman BE, Mark C et al (1984) Tumorigenicity of humon HT1080 fibrosarcoma X normal fibroblast hybrids: chromosome dosage dependency. Cancer Res 44:3471–3479

Bishop JM (1987) The molecular genetics of cancer. Science 235:305–311

Bradbury LE, Kansas GS, Levy S et al (1992) The CD19/CD21 signal transducing complex of human B lymphocytes includes the target of antiproliferative antibody-1 and Leu-13 antigen. J Immunol 149:2841–2850

Bremmer R, Balmain A (1990) Genetic changes in skin tumor progression: correlation between presence of a mutant ras gene and loss of heterozygosity on mouse chromosome 7. Cell 61:407–417

Call KM, Glaser T, Ito CY et al (1990) Isolation and characterization of a zinc finger polypeptide gene at the human chromosome 11 Wilms' tumor locus. Cell 60:509–520

Chen P, Ellmore N, Weissman BE et al (1994) Functional evidence for a second tumor suppressor gene on human chromosome 17. Mol Cell Biol 14:534–542

Conway K, Morgan D, Phillips K et al (1992) Tumorigenic suppression of a human cutaneous squamous cell carcinoma cell line in the nude mouse skin graft assay. Cancer Res 52:6487–6495

Crawford LV (1983) The 53000-dalton cellular protein and its role in transformation. Int Rev Exp Pathol 25:1–50

Dear TN, Kefford RF (1990) Molecular oncogenetics of metastasis. Mol Aspects Med 11:243–324

Devilee P, Van Den Broek M, Mannens M et al (1991) Differences in patterns of allelic loss between two common types of adult cancer, breast and colon carcinoma, and Wilms' tumor of childhood. Int J Cancer 47:817–821

Dong JT, Lang PW, Rinker-Schaeffer CW et al (1995) KAI1, a metastasis suppressor gene for prostate cancer on human chromosome 11p11.2. Science 268:884–886

Dowdy SF, Fasching CL, Scanlon DJ et al (1991) Suppression of tumorigenicity in Wilms' tumor by the p14:p15 region of chromosome 11. Science 254:293–295

Drummond IA, Madden Sl et al (1992) Repression of the insulin-like growth factor II gene by the Wilms' tumor suppressor WT1. Science 257:674–678

Ege T, Ringertz NR (1974) Preparation of microcells by enucleation of micronucleated cells. Exp Cell Res 87:378–382

Eliyahu D, Michalovitz D et al (1989) Wild-type p53 can inhibit oncogene-mediated focus formation. Proc Natl Acad Sci USA 86:8763–8767

Fearon ER, Cho KR, Nigro JM et al (1990) Identification of a chromosome 18q gene that is altered in colorectal cancers. Science 247:49–56

Finlay CA, Hinds PW, Levine AJ (1989) The p53 proto-oncogene can act as a suppressor of transformation. Cell 57:1082–1093

Fournier REK, Ruddle FH (1977) Microcell-mediated transfer of murine chromosomes into mouse, Chinese hamster, and human somatic cells. Proc Natl Acad Sci USA 74:319–323

Friend SH, Bernards S, Rogelj S et al (1986) A human DNA segment with properties of the gene that predisposes to retinoblastoma and osteosarcoma. Nature 323:643–646

Fukudome K, Fururse M, Imai T et al (1992) Identification of membrane antigen C33 recognized by monoclonal antibodies inhibitory to human T-cell leukemia virus type 1 (HTLV-1)-induced syncytium formation: altered glycosylation of C33 antigen in HTLV-1-positive T cells. J Virol 66:1394–1401

Fung Y-K, Murphree Al, Tang A et al (1987) Structural evidence for the authenticity of the human retinoblastoma gene. Science 236:1657–1661

Fusenig NE, Breitkreutz D, Dzarlieva RT, Boukamp P, Bohnert A, Tilgen W (1983) Growth and differentiation characteristics of transformed keratinocytes from mouse and human skin in vitro and in vivo. J Invest Dermatol 81:168s–175s

Futreal PA,Liu Q, Shattuck-Eidens D et al (1994) BRCA1 mutations in primary breast and ovarian carcinomas. Science 266:120–122

Geiser AG, Anderson MJ, Stanbridge EJ et al (1989) Suppression of tumorigenicity in human cell hybrids derived from cell lines expressing different activated ras oncogenes. Cancer Res 49:1572–1577

Gessler M, Poustka A et al (1990) Homozygous deletion in Wilms' tumors of a zinc-finger gene identified by chromosome jumping. Nature 343:774–778

Giard DJ, Aaronson SA, Todaro GJ et al (1974) In vitro cultivation of human tumors: establishment of cell lines derived from a series of solid tumors. J Natl Cancer Inst 51:1417–1423

Gioeli D, Conway K, Weissman BE et al (1997) Localization and characterization of a chromosome 11 tumor suppressor gene using organotypic raft cultures. Cancer Res 57:1157–1165

Groden J, Thilveris A, Samowitz W et al (1991) Identification and characterization of the familiar adenomatous polyposis coli gene. Cell 66:589–600

Hahn H, Wicking C, Zaphiropoulos PG et al (1996a) Mutations of the human homolog of Drosophila patched in the nevoid basal cell carcinoma syndrome. Cell 85:841–851

Hahn SA, Schutte M, Hoque ATMS et al (1996b) DPC4, a candidate tumor suppressor gene at human chromosome 18q21.1. Science 271:350–353

Hampton GM, Mannerma A, Winquist R et al (1994a) Loss of heterozygosity in sporadic human breast carcinoma: a common region between 11q22 and 11q23.3. Cancer Res 54:4586–4589

Hampton GM, Penny LA, Baorgen RN et al (1994b) Loss of heterozygosity in cervical carcinoma: subchromosomal localization of a putative tumor-suppressor gene to chromosome 11q22-q24. Proc Natl Acad Sci USA 91:6953–6957

Hao Y, Crenshaw T, Moulton T et al (1993) Tumor-suppressor activity of H19 RNA. Nature 365:764–767

Harris H, Miller OJ, Klein G et al (1969) Suppression of malignancy by cell fusion. Nature 223:363–368

Hartwell L (1992) Defects in a cell cycle checkpoint may be responsible for the genomic instability of cancer cells. Cell 71:543–546

Helin K, Harlow E (1993) The retinoblastoma protein as a transcriptional repressor. Trends Cell Biol 3:43–46

Heo DS, Snyderman C, Gollin SM et al (1989) Biology, cytogenetics and sensitivity to immunological effector cells of new head and neck squamous cell carcinoma lines. Cancer Res 49:5167–5175

Huang Y, Meltzer SJ et al (1993) Altered messenger RNA and unique mutatinal profiles of p53 and Rb in human esophageal carcinomas. Cancer Res 53:1889–1894

Imai T, Fukudome K, Tagai S et al (1992) C33 antigen recognized by monoclonal antibodies inhibitory to human T cell leukemia virus type 1-induced syncytium formation is a member of a new family of transmembrane proteins including CD9, CD37, CD53, and CD63. J Immunol 149:2879–2886

Johnson RL, Rothman Al, Xie J et al (1996) Human homolog of patched, a candidate gene for the basal cell nevus syndrome. Science 272:1668–1671

Kinzler KW, Nilbert MC, Su L-K, et al (1991a) Identification of FAP locus genes from chromosome 5q21. Science 253:661–669

Kinzler KW, Nilbert MC, Vogelstein B et al (1991b) Identification of a gene located at chromosome 5q21 that is mutated in colorectal cancers. Science 251:1366–1370

Koi M, Morita H, Yamada H et al (1989) Normal human chromosome 11 suppresses tumorigenicity of human cervical tumor cell line SiHa. Mol. Carcinogenesis 2:12–21

Kolodner RD (1996) Biochemistry and genetics of eukaryotic mismatch repair. Genes Dev 10:1433–1442

Koufos A, Grundy P et al (1989) Familial Wiedemann-Beckwith syndrome and a second Wilms' tumor locus both map to 11p15.5. Am J Hum Genet 44:711–719

Kratzke RA, Greatens TM, Rubins JB et al (1996) Rb and p16 INK4a Expression in resected non-small cell lung tumors. Cancer Res 56:3415–3420

Latif F, Tory K, Gnarra J et al (1993) Identification of the von Hippel-Lindau disease tumor suppressor gene. Science 260:1320–1357

Lee W-H, Bookstein R, Hong F et al (1987) Human retinoblastoma susceptibility gene: cloning, identification, and sequence. Science 235:1394–1399

Li J, Yen C, Liaw D et al (1997) PTEN, a putative protein tyrosine phosphatase gene mutated in human brain, breast and prostate cancer. Science 275:1943–1947

Loh WE, Scrable HJ, Livanos E et al (1992) Human chromosome 11 contains two different growth suppressor genes for embryonal rhabdomyosarcoma. Proc Natl Acad Sci USA 89:1755–1759

Miki Y, Swensen J, Shattuck-Eidens D et al (1994) A strong candidate for the breast and ovarian cancer susceptibility gene BRCA1. Science 266:66–71

Misra BC, Srivatsan ES (1989) Localization of HeLa cell tumor-suppressor gene to the long arm of chromosome 11. Am J Human Genet 45:565–577

Muktar H, Bickers DR (1993) Environmental skin cancer: mechanisms, models and human cancer. Cancer Res 53:3439–3442

Negrini M, Rasio D, Hampton GM et al (1992) Suppression of tumorigenesis by the breast cancer cell line MCF-7 following transfer of normal human chromosome 11. Oncogene 7:2013–2018

Negrini M, Castagnoli A, Sabbioni S et al (1994) Suppression of tumorigenicity of breast cancer cells by microcell-mediated chromosome transfer: studies on chromosomes 6 and 11. Cancer Res 54:1331–1336

Negrini M, Sabbioni S, Possati L et al (1995) Definition and refinement of chromosome 11 regions of loss of heterozygosity in breast cancer: identification of a new region at 11q23.3. Cancer Res 55:3003–3007

Nigro JM, Baker SJ, Preisinger AC et al (1989) Mutations in the p53 gene occur in diverse human tumor types. Nature 342:705–708

Nojima Y, Hirose T, Tachibana K et al (1993) The 4F9 antigen is a member of the tetraspan transmembrane protein family and functions as an accessory molecule in T cell activation and adhesion. Cell Immunol 152:249–260

Pasquale SR, Jones GR, Doersen C-J et al (1988) Tumorigenicity and oncogene expression in pediatric cancers. Cancer Res 48:2715–2719

Pereira-Smith OM, Smith JR (1983) Evidence for the recessive nature of cellular immortality. Science 221:964–966

Phillips KK, Welch DR, Miele ME et al (1996) Suppression of MDA-MB-435 breast carcinoma cell metastasis following the introduction of human chromosome 11. Cancer Res 56:1222–1227

Phillips KK, White AE, Hicks DJ et al (1998) Correlation between reduction of metastasis in the MDA-MB-435 model system and increased expression of the Kai-1 protein. Mol Carcinog 21:111–120

Pietenpol JA, Stein RW, Moran E et al (1990) TGF-beta 1 inhibition of c-myc transcription and growth in keratinocytes is abrogated by viral transforming proteins with pRB binding domains. Cell 61:777–785

Qazilbash MH, Xiao X, Seth P et al (1997) Cancer gene therapy using a novel adeno-associated virus vector expressing human wild-type p53. Gene Ther 4:675–682

Reed AL, Califano J, Cairns P et al (1996) High frequency of p16 (CDKN2/MTS-1/INK4 A) inactivation in head and neck squamous cell carcinoma. Cancer Res 56:3630–3633

Reeve AE, Sih SA, Raizis AM et al (1989) Loss of allelic heterozygosity at a second locus on chromosome 11 in sporadic Wilms' tumor cells. Mol Cell Biol 9:1799–1803

Reid LH, West A, Gioli DG et al (1996) Localization of a tumor suppressor gene in 11p15.5 using the G401 Wilms' tumor assay. Hum Mol Genet 5:239–247

Rotter V, Foord O, Navot N (1993) In search of the functions of normal p53 protein. Trends Cell Biol 3:43–46

Savitsky K, Bar-Shira A, Gilad S et al (1995) A single ataxia telangeictasia gene with a product similar to PI-3 kinase. Science 268:1749–1753

Saxon PJ, Srivatsan ES, Stanbridge EJ (1986) Introduction of human chromosome 11 via microcell transfer controls tumorigenic expression of HeLa cells. EMBO J 5:3461–3466

Sherr CJ (1996) Cancer cell cycles. Science 274:1672–1677

Stanbridge EJ (1992) Functional evidence for human tumor suppressor genes: chromosomal and molecular genetic studies. Cancer Surv 12:5–24

Stanbridge EJ, Flandemeyer R, Daniels D et al (1981) Specific chromosome loss associated with the expression of tumorigenicity in human cell hybrids. Somat Cell Genet 7:699–712

Steck PA, Pershouse MA, Jasser SA et al (1997) Identification of a candidate tumor suppressor gene, MMAC1, at chromosome 10q23.3 that is mutated in multiple advanced cancers. Nat Genet 15:356–362

Steeg PS, Bevilacqua G, Pozzatti R et al (1988) Altered expression of NM23, a gene associated with low tumor metastic potential, during adenovirus 2 Ela inhibition of experimental metastasis. Cancer Res 48:6550–6554

Takahashi S, Doss C, Levy S et al (1990) Tapa-1, the target of an antiproliferative antibody, is associated on the cell surface with the Leu-13 antigen. J Immunol 145:2207–2213

Tlsty T, White A, Sanchez J (1992) Suppression of gene amplification in human cell hybrids. Science 255:1425–1427

Trent J, Yang JM, Emerson J et al (1993) Clonal chromosome abnormalities in human breast carcinomas: thirty-four cases with metastatic disease. Genes Chromos Cancer 7:194–203

Trofatter JA, MacCollin MM, Rutter JL et al (1993) A novel Moesin-Ezrin-Radixin-like gene is a candidate for the neurofibromatosis 2 tumor suppressor gene. Cell 72:791–800

Virtaneva KI, Emi N, Marken JS et al (1994) Chromosomal localization of three human genes coding for A15, L6, and S5.7 (TAPA1): all members of the transmembrane 4 superfamily of proteins. Immunogenetics 39:329–334

Viskochil D, Buchberg AM, Xu G et al (1990) Deletions and a translocation interrupt a cloned at the neurofibromatosis type 1 locus. Cell 62:187–192

Vogelstein B, Fearon ER, Kern SE et al (1989) Allelotype of colorectal carcinomas. Science 244:207–211

Vogelstein B, Kinzler KW (1992) p53 function and dysfunction. Cell 70:523–529

Wallace MR, Marchuk DA, Anderson LB et al (1990) Type 1 neurofibromatosis gene: identification of a large transcript disrupted in three NF1 patients. Science 249:181–186

Wang Z, Madden SL, Deuel TF et al (1992) The Wilms' tumor gene product, WT1 represses transcription of the platelet-derived growth factor A-chain gene. J Biol Chem 267:21999–22002

Weinberg RA (1989) The molecular basis of retinoblastomas. Ciba Found Symp 142:99–105

Weissman BE (1990) Genetic behaviour of tumor genicity in human cancer. In: Cavenee W, Ponder B, Solomon E (eds) Cancer surveys-genetics and cancer, vol 9. Oxford University Press, Oxford, pp 475–485

Weissman BE, Conway K (1995) Genetic aspects of tumor suppressor genes. Adv Genome Biol 3A:137–156

Weissman BE, Saxon PJ, Pasquale SR et al (1987) Introduction of a normal human chromosome 11 into a Wilms' tumor cell line controls its tumorigenic expression. Science 236:175–180

Weissman BE, Stanbridge EJ (1983) Complementation of the tumorigenic phenotype in human cell hybrids. J Natl Cancer Inst 70:666–672

Wilson JL, Dollard SC, Chow LT, Broker TR (1992) Epithelial-specific gene expression during differentiation of stratified primary human keratinocyte cultures. Cell Growth Differ 3:471–483

Winquist R, Mannerma A, Alvaikko M et al (1993) Refinement of regional loss of heterozygosity for chromosome 11p15.5 in human breast tumors. Cancer Res 53:4486–4488

Wooster R, Bignell G, Lancaster J et al (1995) Identification of the breast cancer susceptibility gene BRCA2. Nature 378:789–792

Wright MD, Tomlinson MG (1994) The ins and outs of the transmembrane 4 superfamily. Immunol Today 15:588–594

Xiong Y, Zhang H, Beach D (1992) D type Cyclins associated with multiple protein kinases and the DNA replication and repair factor PCNA. Cell 71:505–514

Yamada H, Wake N, Fujimoto S et al (1990) Multiple chromosomes carrying tumor suppressor activity for a uterine endometrial carcinoma cell line identified by microcell-mediated chromosome transfer. Oncogene 5:1141–1147

Yang X, Welch DR, Philips KK et al (1997) KaII, a putative marker for metastatic potential in human breast cancer. Cancer Lett 119:149–155

Zhang H, Tombline G, Weber WL et al (1998) BRCA1, BRCA2, and DNA damage response: collision or collusion? Cell 92:433–436

Zutter MM, Cantoro SA, Stotz WD et al (1995) Re-expression of the $\alpha 2\beta 1$ integrin abrogates the malignant phenotype of breast carcinoma cells. Proc Natl Acad Sci USA 92:7411–7415

5 Oligonucleotide Therapeutics for Human Leukemia

A.M. Gewirtz

5.1 Introduction

As we approach the new millennium, a quick glance backwards reveals that truly astounding progress has been made in the identification of genes responsible for cell growth, development, and neoplastic transformation. With this knowledge has come a natural desire to translate this information into new therapeutic strategies for many of the common maladies which afflict humankind. These include in particular, but are not limited to, cardiovascular, infectious, neurologic, and neoplastic diseases. One obvious way to exploit the insights which have been made in understanding cellular molecular biology is to insert genes into cells which either replace, or counter, the effects of mutated and/or dysfunctional disease causing genes. This technically complex, as yet largely unrealized endeavor (Crystal 1995; Verma 1997), is what most individuals think of when the terms "gene therapy" or "molecular medicine" are discussed. Nevertheless, alternative strategies for treating diseases at the gene level are being developed. The common goal of these various strategies, which are turning out to be as technically demanding as more traditional gene therapy, is to identify disease causing, or disease related, genes and target them for "silencing." Since the numbers of maladies which might be treated by this approach is genuinely enormous, this is clearly a most important field of endeavor.

For the past several years, we have been engaged in trying to develop a effective strategy of disrupting specific gene function with antisense oligodeoxynucleotides (ODNs). We have also been actively engaged in attempting to utilize this strategy in the clinic. This latter pursuit has focused on finding appropriate gene targets that can be successfully targeted using an antisense approach, and then developing "scale-up" methods so that techniques developed in the laboratory could be applied in the clinic. It was our opinion that human leukemias would be particularly amenable to this therapeutic strategy. They can be successfully manipulated ex vivo, the tumor is "liquid" in vivo and therefore more likely to successfully take up ODNs, and a great deal is known about their cell and molecular biology. The latter in particular facilitates choice of a gene target. Accordingly, if ODNs were going to be developed as therapeutics, the hematopoietic system seemed an ideal model system.

5.2 Potential Gene Targets for Treating Human Leukemia

5.2.1 Transcription Factors

Of the genes that we have targeted for disruption using the antisense ODN strategy (Gewirtz and Calabretta 1988; Ratajczak et al. 1992c; Takeshita et al. 1993; Small et al. 1994; Luger et al. 1996) one that has

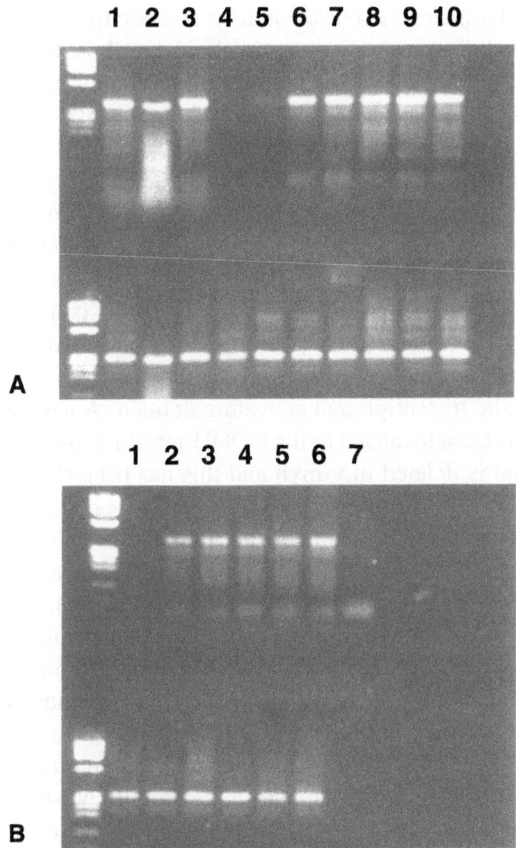

Fig. 1A, B. Vav mRNA expression in erythroid (**A**) and myeloid (**B**) colony cells. Vav mRNA expression was undetectable in cells exposed to the Vav antisense oligodeoxynucleotides (ODNs) at the time cells were first seeded into the cultures (**A**, *lane 4*). Vav mRNA was also undetectable in antisense treated cells after 3 days of growth in methylcellulose (**A**, *lane 5*; **B**, *lane 1*). By day 6, however, Vav mRNA expression was again detectable and remained so on day 8 (**A**, *lanes 6, 7*; **B**, *lanes 2, 3*). β-actin mRNA expression was detectable in these same cells at all time points tested (**A**, **B**, *lower lanes*). Cells exposed to Vav Scr ODNs were assayed for Vav and β-actin mRNA expression as well. When assayed at the same time points, no effect on Vav mRNA expression could be discerned (**A**, *lanes 3, 8–10*; **B**, *lanes 4–6*)

been of particular interest to our laboratory, and one for which therapeutically motivated disruptions are now in clinical trial, is the c-myb gene (Lyon et al. 1994) . C-myb is the normal cellular homologue of v-myb, the transforming oncogene of the avian myeloblastosis virus (AMV) and avian leukemia virus E26. It is a member of a family composed of at least two other highly homologous genes designated A-myb and B-myb (Nomura et al. 1993). Located on chromosome 6q in humans, c-myb's predominant transcript encodes an pprox75 kDa nuclear binding protein (Myb) which recognizes the core consensus sequence 5'-PyAAC(G/Py)G-3' (Biedenkapp et al. 1988). Myb consists of three primary functional regions (Sakura et al. 1989; Fig. 1). At the NH_2-terminal is the DNA binding domain. The mid-portion of the protein contains an acidic transcriptional activating domain. A negative regulatory domain has been localized to the COOH-terminal. Interestingly, the COOH-terminal is deleted in v-myb and this has been thought to contribute to v-myb's transforming ability. Recently reported experiments have been among those to confirm this hypothesis and have further demonstrated that NH_2-terminal deletions give rise to a protein with even more potent transforming ability (Dini et al. 1995). Deletions of both the NH_2- and COOH-terminals create a protein with the greatest transforming ability and one which induces the formation of hematopoietic cells that are more primitive than those produced by NH_2-terminal deletions alone (Dini et al. 1995). These data suggest that simultaneous loss of Myb's ability to bind DNA and interact with as yet unidentified proteins are potent transforming stimuli. Nevertheless, this simple hypothesis is complicated by the observation that overexpression of the COOH-terminal portion of c-myb can also be oncogenic (Press et al. 1994) whereas overexpression of the whole protein is not (Dini et al. 1995). At the least, one may conclude that sequestration of certain potential Myb binding proteins may also be an oncogenic event (Kanei-Ishii et al. 1992). Accordingly, Myb is clearly an important hematopoietic cell gene which may, directly or indirectly, contribute to the pathogenesis or maintenance of human leukemias. For this reason it is a rational target for therapeutically motivated disruption strategies.

5.2.2 Receptors and Signaling Proteins

Other potential mRNA targets for treating leukemia might also be envisioned. These may be found among the surface receptors and signaling proteins constitutive to hematopoietic cells. For example, we have previously suggested that the Kit receptor might serve this purpose (Ratajczak et al. 1992a–c). As noted above, Kit appears to be a c-myb regulated gene, and down-regulation of Kit in hematopoietic cells is associated with induction of apoptosis, perhaps an important mechanistic component of anti-Myb's ability to kill malignant cells. Directly comparing the ability of anti-Myb and anti-Kit targeted oligonucleotides to inhibit the growth of leukemic cells suggests this possibility, while at the same time supports the potential of Kit targeted oligonucleotides in the treatment of chronic myelogenous leukemia (CML) and other leukemias.

A newer target which may prove to be of even greater utility in patients with hematologic malignancies is the protein encoded by the vav proto-oncogene (Bonnefoy-Berard 1996; Katzav 1992). Vav is expressed exclusively in hematopoietic cells where it is assumed to play a role in signaling. The importance of this role is uncertain, in part because of conflicting functional studies which have employed different strategies for abrogating Vav gene expression in murine embryonic stem cells (Wulf 1993; Zhang et al. 1994; Zmuidzinas et al. 1995; Zhang 1995). Our studies with vav targeted oligonucleotides (see below) lend some support to each of the conflicting reports cited above and imply that vav would be a therapeutically attractive target in CML (Luger et al. 1996). Briefly, we have found that while vav appears to be required for erythropoiesis in both normal and malignant hematopoietic cells, malignant myeloid cell growth, in particular myeloid cells derived from CML patients, does appear to be dependent on Vav expression (Luger et al. 1996). The rational for Vav as an antisense target is therefore anchored in the fact that it is differentially used in normal and malignant cells.

5.3 Perturbing Candidate Gene Expression with Oligodeoxynucleotides

5.3.1 C-myb as Target

Our investigations were initially designed to elucidate the role of Myb protein in regulating hematopoietic cell development. Because the results obtained from these studies had obvious clinical relevance, more translationally oriented studies were also undertaken. These have now culminated in clinical trials which are presently ongoing at the Hospital of the University of Pennsylvania. The steps carried out in the clinical development of the c-myb targeted antisense ODNs are summarized below. I will also allude briefly to our initial clinical experience with the myb targeted ODNs.

5.4 In Vitro Experience in the Hematopoietic Cell System

Attempts to exploit the c-myb gene as a therapeutic target for antisense ODNs began as an outgrowth of studies which seeked to define the role of Myb protein in regulating normal human hematopoiesis (Gewirtz and Calabretta 1988; Gewirtz et al. 1989). During the course of these studies it was determined that exposing normal bone marrow mononuclear cells (MNCs) to c-myb antisense ODNs resulted in a decrease in cloning efficiency and progenitor cell proliferation. The effect was lineage indifferent since c-myb antisense DNA inhibited granulocyte-macrophage colony forming units (CFU-GM), CFU-E (erythroid), and CFU-Meg (megakaryocyte). In contrast, c-myb ODNs with the corresponding sense sequence had no consistent effect on hematopoietic colony formation when compared to growth in control cultures. Finally, inhibition of colony formation was also dose related. Inhibition of the targeted mRNA was also demonstrated. Sequence specific, dose related biologic effects accompanied by a specific decrease or total elimination of the targeted mRNA were strong pieces of evidence to suggest that the effects we were observing were due to an antisense mechanism. It should be added that the effects we observed were largely confirmed using homologous recombination (Mucenski et al. 1991).

Since the c-myb antisense ODNs inhibited normal cell growth, we were also interested in determining their effect on leukemic cell growth. While one could reasonably postulate that aberrant c-myb expression or Myb function might play a role in carcinogenesis, demonstrating this was another matter. To address this question, we employed a variety of leukemic cell lines, including those of myeloid and lymphoid origin (Anfossi et al. 1989; Gewirtz et al. 1989). In addition, we also employed primary patient material (Calabretta et al. 1991). The results of these studies suggested that normal and leukemic cells were growth inhibited to different degrees by the myb targeted ODNs and that the leukemic cells were more sensitive. This supposition was supported by additional studies which also suggested a role of the myb targeted ODNs, a potential bone marrow purging agent, in particular for the treatment of CML (Ratajczak et al. 1992a).

5.5 In Vivo Experience: Development of Animal Models

The studies described above were carried out primarily with unmodified DNA. Since we could not give this material to patients we established a human leukemia/SCID mouse model system with which to test the effectiveness of a phosphorothioate compound against in vivo disease (Ratajczak et al. 1992b). To carry out these experiments, SCID mice were injected IV with K562 chronic myeloid leukemia cells after cyclophosphamide conditioning. K562 cells express c-myb, the antisense ODN target, and the tumor specific bcr\abl oncogene which was utilized for tracking the human leukemia cells in the mouse host. After tumor cell injection, animals developed blasts in the peripheral blood within 4–6 weeks. After peripheral blood blast cells appeared, the mean (±SD) survival of untreated mice (n=20) was 6±3 days. Dying animals had prominent central nervous system (CNS) infiltration, marked infiltration of the ovary, and scattered abdominal granulocytic sarcomas. Infusion of either sense (S) or scrambled (Scr) sequence c-myb phosphorothioate ODNs (24 bp; codons 2–9) for 3 7, or 14 days had no statistically significant effect on sites of disease involvement or animal survival in comparison to control animals. In contrast, animals treated for 7 or 14 days with c-myb antisense ODNs survived 3.5 to eight times longer (p0.001) than the various control animals (n=60). In addition, animals receiving c-myb antisense DNA had either rare microscopic foci or no obviously detectable CNS disease, and a 50% reduction of ovarian involvement. These results suggested that phosphorothioate modified c-myb antisense DNA might be efficacious for the treatment of human leukemia in vivo.

5.5.1 Proto-vav as Oligodeoxynucleotide Target

5.5.1.1 Effect of Pulse Oligodeoxynucleotide Exposure on Vav mRNA Expression Kinetics in Cultured Bone Marrow Mononuclear Cells

As mentioned briefly above, we found that vav appears to be required for erythropoiesis in both normal and malignant hematopoietic cells, but that malignant myeloid cell growth, in particular myeloid cells derived from CML patients, did not appear to be dependent on Vav expression (Luger et al. 1996). These experiments were not informative however in defining when Vav function might be most critical for colony development. We therefore determined the effect of Vav ODN exposure on Vav mRNA expression kinetics in normal MNCs cultured in suspension and in methylcellulose. To carry out these experiments, MNCs were first exposed to antisense or control sequence ODNs for 24 h in suspension culture. After 24 h the treated cells were either maintained in suspension culture or plated into methylcellulose medium designed to support either myeloid or erythroid colony formation. Under either of these conditions, cells were accessible for Vav mRNA analysis. MNCs maintained in suspension culture were analyzed at 24, 36, 48, and 96 h after initial addition of ODNs (data not shown). Vav mRNA levels were unchanged in control cells and cells exposed to control ODN (S or Scr). In cells exposed to the antisense ODNs, Vav mRNA was variably detectable at 24 h after initial addition, undetectable at 36 and 48 h, and detectable once more at 96 h; β-actin expression was detectable in all of these samples at all of the time points sampled.

MNCs plated into methylcellulose cultures after ODN exposure were sampled at days 3, 5, and 7 after initial seeding. Regardless of the growth factors added, i.e., regardless of whether the cultures were designed to support erythroid (Fig. 1A) or myeloid (Fig. 1B) colony formation, identical results were obtained. Vav mRNA expression was undetectable in cells exposed to the Vav antisense ODN at the time cells were first seeded into the cultures (Fig.1A, lane 4). Vav mRNA was also undetectable in antisense treated cells after 3 days of growth in methylcellulose (Fig. 1A, lane 5; Fig. 1B, lane 1). By day 6, however, Vav mRNA expression was again detectable and remained so on day 8 (Fig. 1 A, lanes 6, 7, Fig. 1B, lanes 2, 3). β-actin mRNA expression was detectable in these same cells at all time points tested (Fig. 1 A ,B, lower lanes). Cells from colonies exposed to Vav Scr ODN were assayed for Vav and β-actin mRNA expression as well. When assayed at the same time points no effects on Vav mRNA expression could be discerned (Fig. 1A, lanes 3, 8–10, Fig. 1B, lanes 4–6). Accordingly, these results document that Vav mRNA was diminished for at least 24 h in cells undergoing myeloid and erythroid differentiation. They further document that perturbation in Vav mRNA expression was sequence specific. These findings strongly suggest that transient inhibition of Vav mRNA expression selectively inhibits development of cells committing to erythroid, as opposed to myeloid, cell growth. Whether continuous inhibition of Vav mRNA expression would further inhibit erythroid cell development is unknown.

Finally, and of obvious importance, to confirm that loss of mRNA expression was accompanied by a decrease in the targeted protein, we also carried out western blotting on progeny of CD34+ MNCs which had been exposed to either antisense or Scr Vav ODNs and then cultured in methylcellulose as described above. On day 3 or day 7 of culture, cells were washed, suspended in Laemmli buffer, and the lysate corresponding to 2×10^5 cells was loaded into each lane. Immunoblotting was then carried out as detailed. Figure 2 reveals that immunodetectable p95 Vav was greatly diminished at day 3 in cells exposed to the antisense ODN when compared to cells exposed to the Scr sequence. This visual impression was confirmed by densitometric scanning which revealed an approx75% decrease in the antisense band density vs the Scr band density (27 vs 102 arbitrary density units, respectively). The identities of the slightly heavier band, and the lighter band in the day 3 antisense lane are not known but these could represent an alternatively glycosylated and a degradation product respectively. At day 7, when the mRNA kinetic analysis suggests that Vav mRNA levels would be present in the two populations, the immunoblot is also consistent as the protein levels in equal numbers of cells appear equal. The specificity of the antibody is

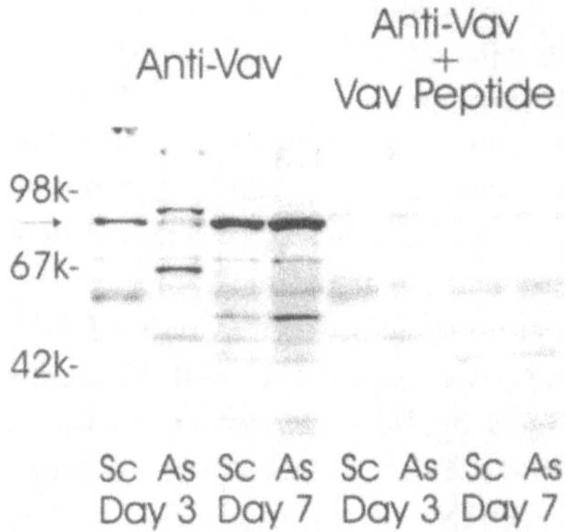

Fig. 2. Western blot of CDC34+ bone marrow mononuclear cells (MNCs) which had been exposed to either antisense (AS) or scrambled (Scr) VAV oligodeoxynucleotides (ODNs) and then cultured in methylcellulose. On day 3 or day 7 of culture, cells were washed, suspended in Laemmli buffer, and the lysate corresponding to 2×10^5 cells was loaded into each lane. Immunodetectable p95 Vav was greatly diminished at day 3 in cells exposed to the AS ODNs compared to cells exposed to the Scr sequence. At day 7, when the mRNA kinetic analysis suggests that Vav mRNA levels would be present in two populations, the immunoblot is also consistent, as the protein levels in equal numbers of cells appear equal. The specificity of the antibody is shown on the right side, where preabsorption of the anti-Vav antibody with a Vav peptide prior to blotting results in disappearance of the p95 band

shown on the right side of the figure, where preabsorption of the anti-Vav antibody with a Vav peptide prior to blotting results in disappearance of the p95 band.

5.5.2 Vav mRNA Expression and Malignant Progenitor Cell Cloning Efficiency

Proto-Vav mRNA transcripts have been detected in virtually all leukemic cell lines tested but the physiologic significance of such expression remains uncertain. To determine the potential importance of Vav expression in primary malignant hematopoietic cells, peripheral blood or bone marrow MNCs were obtained from patients with either

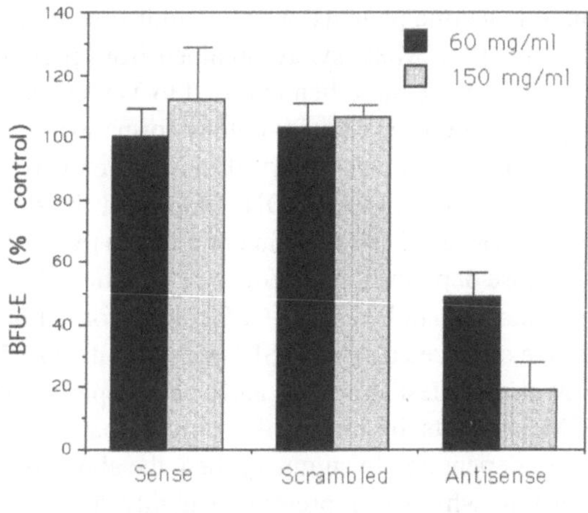

Fig. 3. Effect of oligonucleotides on burst forming unit-erythroid (BFU-E) colony formation from seven patients with polycythemia vera (PV). Bone marrow mononuclear cells were obtained from peripheral blood,partially purified and then exposed to Vav oligodeoxynucleotides (ODNs). At the highest doses of antisense ODNs employed, there was a mean (±SD) decrease of 81%±4% (p.0001) compared to control cultures. As indicated, inhibition was dose dependent and sequence specific

acute or chronic myelogenous leukemia. Depending on the amount of material available from a given patient, MNCs were variably purified (A—,A-T—, CD34+) and then tested for the ability of their CFU to form GM colonies after exposure to Vav ODNs. In distinct contrast to results obtained with normal marrow cells, inhibition of GM colony formation was observed in a variable subset of these patients. This proved to be a minority in the small number of AML patient specimens examined (1 of 3 patients evaluated). In addition, inhibition of colony growth was modest (pprox35%) in the one patient sample in which this was observed. A much more profound effect was observed in the CML patients, 14 of whom were in chronic phase and three in blast crisis. In this population mean inhibition was pprox80% (*p*.001) compared to control cultures and was observed in 13 of the 17 cases. Inhibition was also sequence specific.

In addition to leukemia patients, we also studied seven patients with polycythemia vera (PV). MNC swere obtained from peripheral blood, partially purified as above, and then exposed to Vav ODNs. Effects on BFU-E (burst forming unit-erythroid) colony formation were assessed. Not surprisingly, inhibition was observed in all seven patients studied. At the highest doses of antisense ODNs employed there was a mean (±SD) decrease of 81%±4% (*p*.0001) in comparison to control cultures. This effect was dose dependent and sequence specific (Fig. 3). In addition, we concomitantly studied CFU-GM colony formation in two of these patients. We observed a mean (±SD) decrease of 76%±5% in these two patients at the highest doses of antisense employed (*p*.002). As noted in the CML patients, inhibition of colony formation was sequence specific and dose dependent. In summary then, the above studies should clearly demonstrate why Vav represents a highly desirable target for antisense ODN mediated mRNA destruction. Several complementary lines of evidence suggest that normal myeloid progenitor cells do not require Vav for development (Zhang et al. 1994; Zmuidzinas et al. 1995). Why Vav might be required by CML cells is revealed by the fact that Vav has been shown to be a downstream element in bcr/abl signaling (Matsuguchi et al. 1995). That fact that erythropoiesis may be transiently disrupted by the Vav antisense is clinically of no concern because the half life of circulating red blood cells is 120 days and because the hemoglobin is easily, and safely, supported by blood cell transfusions. Accordingly, Vav is a rational target in CML and perhaps other myeloid leukemias as well.

5.5.3 Use of Antisense Oligonucleotides in a Clinical Setting

For the purpose of developing an antisense oligonucleotide therapeutic, CML seemed to us to be an excellent disease model. As mentioned above, CML is relatively common and it has a convenient marker chromosome and gene for objectively following potential therapeutic efficacy of a test compound (Gale et al. 1993). In addition to these considerations, CML is uniformly fatal except for individuals who are fortunate enough to have an allogeneic bone marrow donor. Picking a gene target in CML was actually somewhat problematic. An obvious target was the bcr/abl gene encoded mRNA (Melo 1996). However, because bcr/abl is not expressed in primitive hematopoietic stem cells (Bedi et al. 1993) and because it is uncertain if transient interruption of bcr/abl signaling actually results in the death of CML cells, we felt that an alternative target might be of greater use in treating this disease. Based on the type of data presented above, a favorable therapeutic index in toxicology testing, and more detailed knowledge of the pharmacokinetics of oligonucleotides, we have begun to evaluate the myb targeted antisense ODNs in the clinic (Gewirtz et al. 1996a).

Towards this end, we initiated clinical trials to evaluate the effectiveness of phosphorothioate modified ODN antisense to the c-myb gene as marrow purging agents for chronic phase (CP) or accelerated phase (AP) CML patients, and a Phase I intravenous infusion study for blast crisis (BC) patients, and patients with other refractory leukemias. ODN purging was carried out for 24 h on CD34+ marrow cells. Patients received busulfan and cytoxan, followed by re-infusion of previously cryopreserved P-ODN purged MNCs. In the pilot marrow purging study seven CP and one AP CML patients were treated; seven of the eight were engrafted. In four out of six evaluable CP patients, metaphases were 85%–100% normal 3 months after engraftment suggesting that a significant purge had taken place in the marrow graft. Five CP patients demonstrated marked, sustained, hematologic improvement with essential normalization of their blood counts. Follow-up ranges from 6 months to pprox2 years. In an attempt to further increase purging efficiency we incubated patient MNCs for 72 h in the phosphorothioate-ODN. Though PCR and LTCIC studies suggested a very efficient purge had occurred, engraftment in five patients was poor. In the Phase I systemic infusion study, 18 refractory leukemia patients (two patients were treated at two different dose levels; 13 had AP or BC CML). Myb antisense ODNs were delivered by continuous infusion at dose levels ranging between 0.3 mg/kg per day for 7 days to 2.0 mg/kg per day for 7 days. No recurrent dose related toxicity has been noted, though idiosyncratic toxicities, not clearly drug related, were observed (1 transient renal insufficiency; 1 pericarditis). One BC patient survived pprox14 months with transient restoration of CP disease. These studies show that ODNs may be administered safely to leukemic patients. Whether patients treated in either study derived clinical benefit is uncertain, but the results of these studies suggest to us that ODNs may eventually demonstrate therapeutic utility in the treatment of human leukemias.

5.5.4 Problem Solving

The power of the antisense approach has been demonstrated in experiments in which critical biological information has been gathered using antisense technology and has been subsequently verified by other laboratories using other methodologies (Gewirtz and Calabretta 1988; Mucenski et al. 1991; Metcalf 1994). However, this technology, in spite of its successes, has been found to be highly variable in its efficiency. To the extent that many have tried to employ ODNs and more than a few have been perplexed and frustrated by results that were noninformative at best, or even worse, misleading or unreproducible, it is easy to understand why this approach has become somewhat controversial. We believe that progress on two fronts would help address this problem.

First, in order for an ODN to hybridize with its mRNA target, it must find an accessible sequence. Sequence accessibility is at least in part a function of mRNA physical structure, which is dictated in turn by internal base composition and associated proteins in the living cell. Attempts to describe the in vivo structure of RNA, in contrast to DNA, have been fraught with difficulty (Baskerville and Ellington 1995). Accordingly, mRNA targeting is largely a hit or miss process, accounting for many experiments in which the addition of an ODN yields no effect on expression. Hence, the ability to determine which regions of a given mRNA molecule are accessible for ODN targeting is a significant impediment to the application of this technique in many cell systems. We have begun to approach this issue by developing a footprinting assay to determine which physical areas of an RNA are accessible to the oligonucleotide (Fig. 4). We have proceeded under the assumption that sequence which remains accessible to single stranded RNases in a more physiologic environment may also remain accessible for hybridization with an ODN. Preliminary experiments performed in our laboratory, in which a labeled RNA transcript is allowing to hybridize with an oligonucleotide in the presence or absence of nuclear extracts from the cells of interest along with RNase T1, suggest that footprinting of this

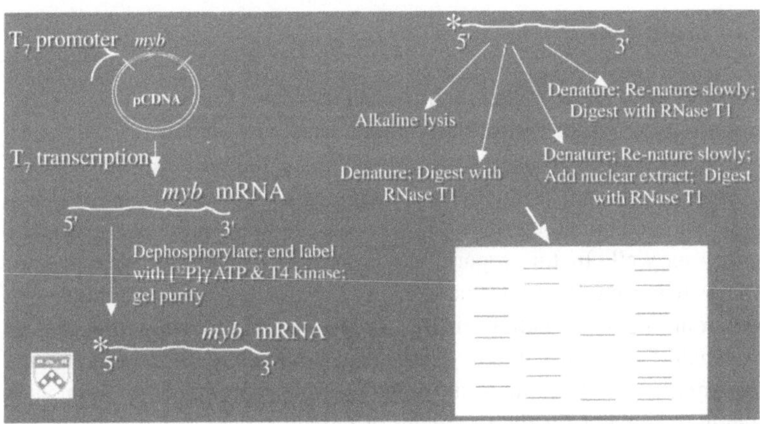

Fig. 4. RNA footprinting methodology for designing oligonucleotides

type is feasible. Of more interest, our preliminary results suggest that this approach may be of use in designing oligonucleotides.

Second, the ability to deliver ODNs into cells and have them reach their target in a bioavailable form also remains problematic (Gewirtz et al. 1996b). Without this ability, it is clear that even an appropriately targeted sequence is not likely to be efficient. Native phosphodiester ODNs, and the widely used phosphorothioate modified ODNs, which contain a single sulfur substituting for oxygen at a non-bridging position at each phosphorus atom, are polyanions. Accordingly, they diffuse across cell membranes poorly and are only taken up by cells through energy dependent mechanisms. This appears to be accomplished primarily through a combination of adsorbtive endocytosis and fluid phase endocytosis which may be triggered in part by the binding of the ODN to receptor-like proteins present on the surface of a wide variety of cells (Loke et al. 1989; Beltinger et al. 1995). After internalization, confocal and electron microscopy studies have indicated that the bulk of the ODNs enter the endosome/lysosome compartment. These vesicular structures may become acidified and acquire other enzymes which degrade the ODNs. Biologic inactivity is the predictable result of this process. Recently described strategies for introducing ODN into cells, including various cationic lipid formulations, may address this problem (Spiller and Tidd 1995; Bergan et al. 1996; Lewis et al. 1996).

5.6 Conclusions

The ability to block gene function with antisense ODNs has become an important tool in many research laboratories. Since activation and aberrant expression of protooncogenes appears to be an important mechanism in malignant transformation, targeted disruption of these genes and other molecular targets with ODNs could have significant therapeutic utility as well. In this regard, the potential therapeutic usefulness of ODNs has been demonstrated in many systems and against a number of different targets including viruses, oncogenes, protooncogenes, and an increasing array of cellular genes. These studies in aggregate suggest that synthetic ODNs have the potential to become an important new therapeutic agent for the treatment of human cancer. Nevertheless, it is clear that considerable optimization will be required before antisense oligonucleotides will emerge as an effective agent for treating human disease. Progress will need to occur on several fronts, including issues related to the chemistry of the molecules employed, for example, how chemical modification impacts uptake, stability, and hybridization efficiency of the synthetic DNA molecule. A clearer understanding of the mechanism of antisense mediated inhibition, including where such inhibition takes place, will also be required. Finally, cellular defense mechanisms such as increasing transcription of the targeted message, may also be factors to consider in planning effective treatment strategies with these agents. It is also straightforward that choice of target is an important consideration. Nevertheless, while many issues remain to be resolved, we remain optimistic that this approach will one day prove useful for the treatment of patients with a variety of hematologic malignancies.

Acknowledgements. Supported by grants from the NIH and the Leukemia Society of America.

References

Anfossi G, Gewirtz AM, Calabretta B (1989) An oligomer complementary to c-myb-encoded mRNA inhibits proliferation of human myeloid leukemia cell lines. Proc Natl Acad Sci USA 86(9):3379–3383

Baskerville S, Ellington AD (1995) RNA structure. Describing the elephant. Curr Biol 5(2):120–123

Bedi A, Zehnbauer BA, Collector MI, Barber JP, Zicha MS, Sharkis SJ, Jones RJ (1993) BCR-ABL gene rearrangement and expression of primitive hematopoietic progenitors in chronic myeloid leukemia. Blood 81(11):2898–2902

Beltinger C, Saragovi HU, Smith RM, LeSauteur L, Shah N, DeDionisio L, Christensen L, Raible A, Jarett L, Gewirtz AM (1995) Binding, uptake, and intracellular trafficking of phosphorothioate-modified oligodeoxynucleotides. J Clin Invest 95(4):1814–1823

Bergan R, Hakim F, Schwartz GN, Kyle E, Cepada R, Szabo JM, Fowler D, Gress R, Neckers L (1996) Electroporation of synthetic oligodeoxynucleotides: a novel technique for ex vivo bone marrow purging. Blood 88(2):731–741

Biedenkapp H, Borgmeyer U, Sippel AE, Klempnauer KH (1988) Viral myb oncogene encodes a sequence-specific DNA-binding activity. Nature 335(6193):835–837

Bonnefoy-Berard N, Munshi A, Yron I, Wu S, Collins TL, Deckert M, Shalom-Barak T, Giampa L, Herbert E, Hernandez J, Meller N, Couture C, Altman A (1996) Vav: function and regulation in hematopoietic cell signaling. Stem Cells 14:250-68

Calabretta B, Sims RB, Valtieri M, Caracciolo D, Szczylik C, Venturelli D, Ratajczak M, Beran M, Gewirtz AM (1991) Normal and leukemic hematopoietic cells manifest differential sensitivity to inhibitory effects of c-myb antisense oligodeoxynucleotides: an in vitro study relevant to bone marrow purging. Proc Natl Acad Sci USA 88(6):2351–2355

Crystal RG (1995) Transfer of genes to humans: early lessons and obstacles to success. Science 270:404-410

Dini PW, Eltman JT et al (1995) Mutations in the DNA-binding and transcriptional activation domains of v-Myb cooperate in transformation. J Virol 69(4):2515–2524

Gale RP, Grosveld G et al (1993) Chronic myelogenous leukemia: biology and therapy. Leukemia 7(4):653–658

Gewirtz AM, Calabretta B (1988) A c-myb antisense oligodeoxynucleotide inhibits normal human hematopoiesis in vitro. Science 242(4883):1303–1306

Gewirtz AM, Anfossi G, Venturelli D, Valpreda S, Sims R, Calabretta B (1989) G1/S transition in normal human T-lymphocytes requires the nuclear protein encoded by c-myb. Science 245(4914):180–183

Gewirtz AM, Luger S, Sokol D, Gowdin B, Stadtmauer E, Reccio A, Ratajczak MZ (1996a) Oligodeoxynucleotide therapeutics for human myelogenous leukemia: interim results. Blood 88 [Suppl 1,10]:270a

Gewirtz AM, Stein CA, Glazer PM (1996b) Facilitating oligonucleotide delivery: helping antisense deliver on its promise. Proc Natl Acad Sci USA 93(8):3161–3363

Kanei-Ishii C, MacMillan EM et al (1992) Transactivation and transformation by Myb are negatively regulated by a leucine-zipper structure. Proc Natl Acad Sci USA 89(7):3088–3092

Katzav S (1992) vav: a molecule for all haemopoiesis? Br J Haematol 81:141-144

Lewis JG, Lin KY, Kothavale A, Flanagan WM, Matteucci MD, DePrince RB, Mook RA, Jr., Hendren RW, Wagner RW (1996) A serum-resistant cytofectin for cellular delivery of antisense oligodeoxynucleotides and plasmid DNA. Proc Natl Acad Sci USA 93(8):3176–3181

Loke SL, Stein CA, Zhang XH, Mori K, Nakanishi M, Subasinghe C, Cohen JS, Neckers LM (1989) Characterization of oligonucleotide transport into living cells. Proc Natl Acad Sci USA 86(10):3474–3478

Luger SM, Ratajczak J, Ratajczak MZ, Kuczynski WI, DiPaola RS, Ngo W, Clevenger CV, Gewirtz AM (1996) A functional analysis of protooncogene Vav's role in adult human hematopoiesis. Blood 87(4):1326–1334

Lyon J, Robinson C et al (1994) The role of Myb proteins in normal and neoplastic cell proliferation. Crit Rev Oncog 5(4):373–388

Matsuguchi, T, Inhorn RC et al (1995) Tyrosine phosphorylation of p95Vav in myeloid cells is regulated by GM-CSF, IL-3 and steel factor and is constitutively increased by p210BCR/ABL. EMBO J 14(2):257–265

Melo JV (1996) The molecular biology of chronic myeloid leukaemia. Leukemia 10(5):751–756

Metcalf D (1994) Blood. Thrombopoietin–at last (news; comment). Nature 369(6481):519–520

Mucenski ML, McLain K, Kier AB, Swerdlow SH, Schreiner CM, Miller TA, Pietryga DW, Scott WJ, Jr., Potter SS (1991) A functional c-myb gene is required for normal murine fetal hepatic hematopoiesis. Cell 65(4):677–689

Nomura N, Zu YL, Maekawa T, Tabata S, Akiyama T, Ishii S (1993) Isolation and characterization of a novel member of the gene family encoding the cAMP response element-binding protein CRE-BP1. J Biol Chem 268(6):4259–4266

Press RD, Reddy EP et al (1994) Overexpression of C-terminally but not N-terminally truncated Myb induces fibrosarcomas: a novel nonhematopoietic target cell for the myb oncogene. Mol Cell Biol 14(4):2278–2290

Ratajczak MZ, Hijiya N et al (1992a) Acute- and chronic-phase chronic myelogenous leukemia colony-forming units are highly sensitive to the growth inhibitory effects of c-myb antisense oligodeoxynucleotides. Blood 79(8):1956–1961

Ratajczak MZ, Kant JA, Luger SM, Hijiya N, Zhang J, Zon G, Gewirtz AM (1992b) In vivo treatment of human leukemia in a scid mouse model with c-myb antisense oligodeoxynucleotides. Proc Natl Acad Sci USA 89(24):11823–11827

Ratajczak MZ, Luger SM, DeRiel K, Abrahm J, Calabretta B, Gewirtz AM (1992c) Role of the KIT protooncogene in normal and malignant human hematopoiesis. Proc Natl Acad Sci USA 89(5):1710–1714

Sakura H, Kanei-Ishii C et al (1989) Delineation of three functional domains of the transcriptional activator encoded by the c-myb protooncogene. Proc Natl Acad Sci USA 86(15):5758–5762

Small D, Levenstein M, Kim E, Carow C, Amin S, Rockwell P, Witte L, Burrow C, Ratajczak MZ, Gewirtz AM et al. (1994) STK-1, the human homolog of Flk-2/Flt-3, is selectively expressed in CD34+ human bone marrow cells and is involved in the proliferation of early progenitor/stem cells. Proc Natl Acad Sci USA 91(2):459–463

Spiller DG, Tidd DM (1995) Nuclear delivery of antisense oligodeoxynucleotides through reversible permeabilization of human leukemia cells with streptolysin O. Antisense Res Dev 5(1):13–21

Takeshita K, Bollekens JA et al. (1993) A homeobox gene of the Antennapedia class is required for human adult erythropoiesis. Proc Natl Acad Sci USA 90(8):3535–3538

Wulf GM, Adra CN, Lim B (1993) Inhibition of hematopoietic development from embryonic stem cells by antisense vav RNA. Embo J 12:5065-74

Zhang R, Tsai FY, Orkin SH (1994) Hematopoietic development of vav-/- mouse embryonic stem cells. Proc Natl Acad Sci USA 91(26):12755–12759

Zhang R, Alt FW, Davidson L, Orkin SH, Swat W (1995) Defective signalling through the T- and B-cell antigen receptors in lymphoid cells lacking the vav proto-oncogene. Nature 374:470-473

Zmuidzinas A, Fischer KD, Lira SA, Forrester L, Bryant S, Bernstein A, Barbacid M (1995) The vav proto-oncogene is required early in embryogenesis but not for hematopoietic development in vitro. EMBO J 14(1):1–11

6 Ribozymes as Biotherapeutic Tools for the Modulation of Gene Expression

B. Anderegg, A. Irie, and K.J. Scanlon

6.1 Introduction

The better the expression patterns and interactions of genes have been understood, the more it has become apparent that carcinogenesis is a multistep process of genetic alterations of oncogenes and/or tumor suppressor genes (Foulds 1958; Fearon and Vogelstein 1990). Small molecules interfering with genetic deregulation might become key players in the future search for new anti-cancer drugs. Examples of such molecules include antisense oligonucleotides (Zamecnik and Stephenson 1978; Melton 1985; Stein and Cheng 1993), triplex forming oligonucleotides (Felsenfeld et al. 1957; Morgan and Wells 1968; Curcio et al. 1997), peptide nucleic acids (Demidov et al. 1993), and aptamers (Stull and Szoka 1995); ribozymes as small RNA molecules with intrinsic specific catalytic activity are one of the most promising candidates for future molecular drugs (Irie et al. 1997).

Theoretically, a specific ribozyme can be designed to target any undesirably expressed mRNA. Herein lies the potential use of ribozymes as therapeutics for neoplastic disorders and viral illnesses, like those associated with human immunodeficiency virus type 1 (HIV-1) or chronic hepatitis B virus, and as tools for conversion of multidrug resistance (MDR). The crucial role of ribozymes would be the inhibition of information transfer from the gene to the protein by interfering with intermediate mRNA function. Although this is the mechanism of action of antisense oligonucleotides as well, some studies have shown that ribozymes might have superior efficacy due to their cleavage potential (Cameron and Jennings 1989; Sioud et al. 1992; Homann et al. 1993; Scanlon et al. 1994).

This chapter discusses the biochemical facts and problems of the different types of ribozymes known today; considerations for the design of the appropriate ribozyme in a given disease and cell type; and the advantages and disadvantages of the various means of transfer into and expression inside of the cell.

6.2 Mechanisms of Action

Originally, ribozymes were discovered within the group I intervening sequence of the pre-rRNA of *Tetrahymena thermophila*. This sequence is called self-splicing because it catalyzes its own excision (Cech et al. 1981; Kruger et al. 1982). The first truly catalytic ribozyme reported was the RNA portion of the enzyme RNase P purified from *Escherichia coli*, which has the capacity for multiple turnovers (Guerrier-Takada et al. 1983). The satellite RNA of the tobacco ring spot virus (TobRV; Buzayan et al. 1986), the avocado sun blotch viroid (Hutchins et al. 1986), the virusoid lucerne transient streak virus (Forster and Symons

1987), the human hepatitis δ virus (HDV; Branch and Robertson 1991), and a ribosomal RNA (Symons 1992) have also been shown to exhibit self-cleaving properties. Initially, ribozymes were thought to cleave in *cis* conformation only. However, Uhlenbeck (1987) demonstrated true catalytic activity in *trans* for hammerhead ribozymes (Bratty et al. 1993), as did Hampel et al. (1990) for hairpin ribozymes. These properties can be used for highly specific inhibition of gene expression by mRNA targeting (Haseloff and Gerlach 1988).

6.2.1 Chemistry

For ribozyme activity, two different mechanisms are postulated: The term "sterile blocking" refers to the inhibition of mRNA translation or metabolism by binding of an antisense compound to the RNA molecule (Melton 1985; Boiziau et al. 1991). Based on the second mechanism, permanent inactivation of the targeted mRNA by cleavage, ribozymes are theoretically superior to antisense molecules. Like enzymes, ribozymes have the potential for multiple turnover of target molecules.

Until now, six catalytically active RNA motifs have been identified: group I introns, RNase P, the axehead motif of the HDV (Branch and Robertson 1991; Perotta and Been 1992), RNA transcripts of the mitochondrial DNA plasmid of *Neurospora*, the hammerhead and the hairpin motif (Symons 1992, 1994). The latter two will be discussed in further detail below. In general, the chemical mechanism by which ribozymes catalyze the cleavage of their target molecules is rearrangement of phosphodiester bonds (van Tol et al. 1990; Kumar and Ellington 1995).

6.2.2 Hammerhead Ribozymes and Derivatives

The hammerhead motif (Fig. 1a) was originally identified in the satellite RNA of TobRV (Buzayan et al. 1986) where it provides efficient self-cleavage during replication (Gerlach et al. 1987). In the presence of divalent cations such as Mg^{2+} or Pb^{2+}, hammerhead ribozymes can also cleave in *trans* through nonhydrolytic transesterification of the substrate (Uhlenbeck 1987; Haseloff and Gerlach 1988). This happens by activa-

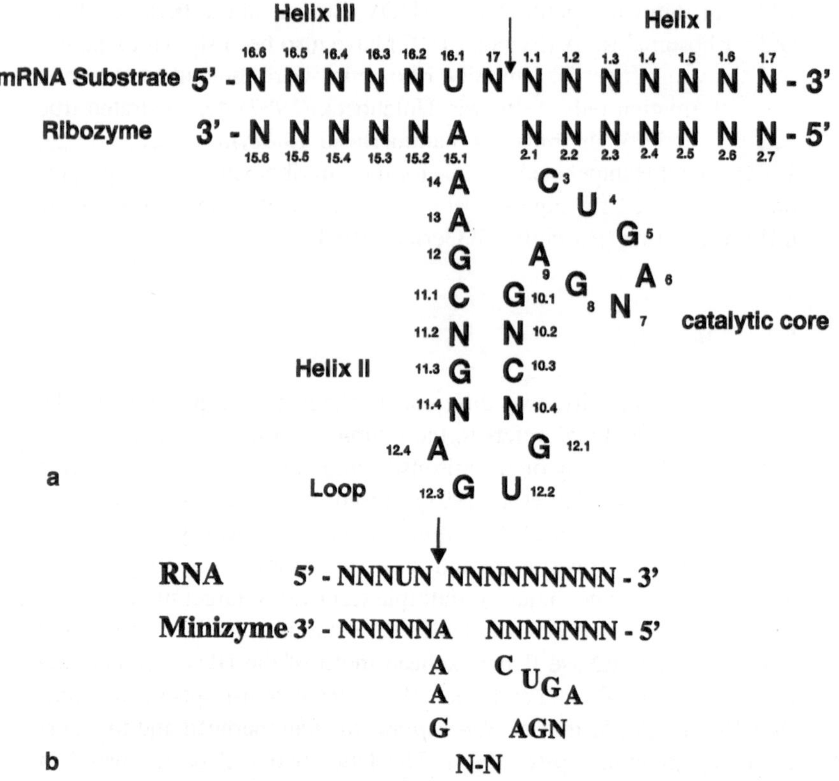

Fig. 1a–c. Synthetic ribozyme structures. **a** Hammerhead ribozyme; **b** minizyme. *N*, any ribonucleotide. The *arrows* indicate the cleavage site within the target RNA (modified from Irie et al. 1997)

tion of the 2'-hydroxyl adjacent to the particular phosphodiester bond. Thereby, a 5'-hydroxyl is displaced and a cyclic 2', 3'-phosphodiester is formed (Buzayan et al. 1988; Pyle 1993; Ohkawa et al. 1995). During viral replication, the multimers newly synthesized by the rolling-circle mechanism are cleaved into monomers by this process.

The secondary structure of the hammerhead ribozyme:substrate complex consists of three helices, a catalytic core region, and a loop sequence. The target is bound through helices I and III, and C^3–A^9 (numbered according to the system published by Hertel et al. 1992) have

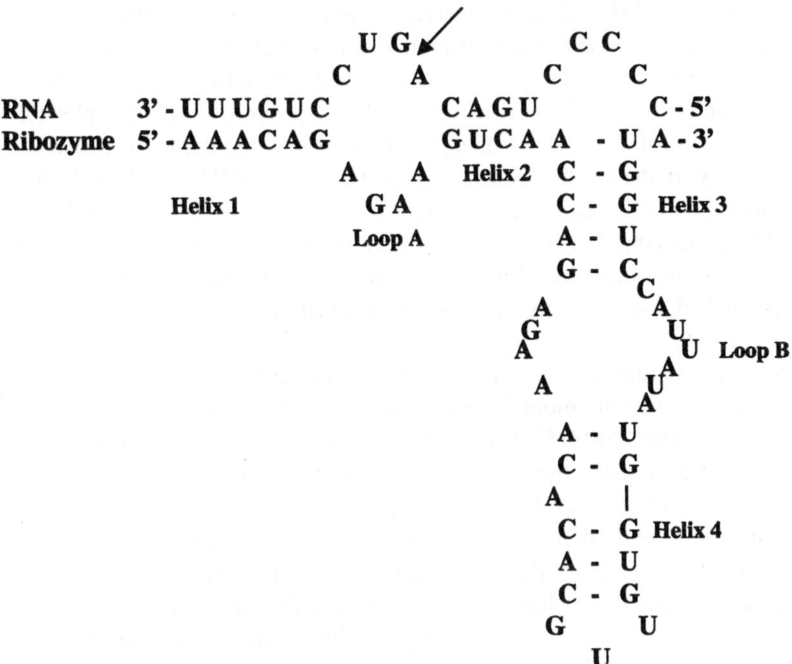

Fig. 1. c Hairpin ribozyme

been shown to form a sharp turn (Sigurdsson et al. 1995; Scott et al. 1995; McKay 1996). This so-called catalytic pocket inside the nonhelical catalytic core with its conserved $C^3U^4G^5A^6$ stretch resembles sequentially and structurally the uridine turn found in the anticodon loop of $tRNA^{Phe}$ (Pley et al. 1994).

Cleavage of the target sequence occurs on the 3' side of the N^{17} residue. Its sequence specificity is high enough to leave RNAs very closely related to the target sequence or even messages with a single base pair mutation unaffected (Bennett and Cullimore 1992; Koizumi et al. 1989). The requirements for activation of the cleaving potential have been studied by several mutational analyses. These demonstrated that the nucleotides C^3–A^6, G^8–$G^{10.1}$, and $C^{11.1}$–A^{14} are essential for the catalytic reaction. 3' of the mRNA cleavage site, a nucleotide triplet at position 16.2, 16.1, and 17 is required, of which the nucleotide at 16.1

has to be a U (Haseloff and Gerlach 1988; Koizumi et al. 1989). In the other two positions of this triplet, it seems that any nucleotide is allowed, although a G at position 17 has been shown to possess very poor cleaving activity (Ruffner et al. 1990). In general, a XUN triplet (X being any nucleotide, N being A, C, or U) is required for cleavage. Usually, if the triplet sequence is GUC, GUA, GUU, CUC, or UUC, cleavage is efficient; if it is GUG, GGC, GAC, CGC, UGC, AUC, or AGC, cleaving efficiency is poor (Perriman et al. 1992). Usually, the AUA is not cleaved, although Nakamaye and Eckstein (1994) have published some activity of a hammerhead ribozyme attacking this triplet.

The in vitro stability and kinetics of hammerhead ribozymes can be altered by various modifications. Nuclease resistance can be increased about 53 000- to 80 000-fold by addition of a 3', 3'-linked thymidine without decreasing the catalytic activity of the ribozyme (Beigelman et al. 1995). In general, the enzymatic kinetics of ribozymes favor short substrates and demonstrate lower efficiency with substrates of 200–1000 base pairs (bp) in length (Heidenreich and Eckstein 1992). Hertel et al. (1994) have found values for the catalytic efficiency of $k_{cat}/K_m=3\times10^5$/M per second using a 17 bp substrate under single turnover conditions. This value is reduced approximately 100-fold when instead of the cleavage step the product release becomes rate limiting, i.e., when the substrate is longer and the secondary structure more complex (Heidenreich et al. 1994). The length and base composition of the flanking sequences (helices I and III) also play an important role in enzyme kinetics. The longer they are, the slower the dissociation step of the ribozyme from the substrate (Herschlag 1991). Bertrand et al. (1994) have reported 12 bp to be optimal. Maximum discrimination is expected to be achieved by AU-rich sequences rather than with GC-rich sequences (Herschlag 1991).

The so-called minizymes (Fig. 1b), interesting derivatives of the hammerhead ribozyme, are capable of overcoming the limiting factor of target accessibility in complexly folded RNA (McCall et al. 1992). Studies with a set of different minizymes suggest that helix II might not be essential for the enzymatic activity. However, it cannot be reduced to less than 2 bp without profound reduction of catalytic activity (Tuschl and Eckstein 1993). Although a minizyme against the TAT transcript of HIV-1 has been found more effective than its full-length counterpart

(Hendry et al. 1995), minizymes generally are expected to be less active than hammerhead ribozymes.

Enhancement of the catalytic efficacy can be achieved by substituting deoxyribonucleotides for ribonucleotides in helices I and III (Hendry et al. 1992). Introduction of deoxyribonucleotides with phosphorothioate linkage in helices I and III and stem II of hammerhead ribozymes not only increased the enzymatic activity, but also nuclease resistance in vivo (Shimayama et al. 1993). Even substitution of all but the ribonucleotide in position 15.1 has shown to have more catalytic efficiency than the all-RNA ribozyme (Chartrand et al. 1995). By contrast, deoxynucleotide substitutes in stem II can lead to marked reduction of cleavage potential (Taylor et al. 1992).

6.2.3 Hairpin Ribozymes

The minus strand of TobRV represents the basic hairpin motif (Fig. 1c). It has the potential to site-specifically *trans*-cleave RNA targets (Hampel and Tritz 1989; Feldstein et al. 1989). Like hammerhead ribozymes hairpin ribozymes cleave their targets in a multistep reaction involving nonhydrolytic transesterification of the substrates (van Tol et al. 1990; Kumar and Ellington 1995). The major difference between these two ribozyme types lies within the strong inhibition of self-cleavage of the hammerhead but not the hairpin by a phosphorothioate modification (Buzayan et al. 1988). In terms of therapeutic application, this might provide a potential advantage of the hammerhead ribozyme.

As suggested by computer modeling, mutational analyses, and phylogenetic studies, the hairpin ribozyme:substrate RNA complex seems to consist of four helical and intervening loop sequences. Binding occurs through the helices 1 and 2, cleavage at the 5' side of a G within loop A of the substrate. Most of the nucleotides in both internal loops are essential for ribozyme activity: In loop A, these are the aforementioned G^6 within the substrate and G^8, A^9, and A^{10} within the ribozyme sequence; in loop B, all nucleotides except A^{20} and U^{39} are essential (Chowrira et al. 1991; Berzal-Herranz et al. 1993). In addition, the 2'-OH groups at A^{10}, G^{11}, A^{24}, and C^{25} cannot be deleted without marked loss of enzymatic activity (Chowrira et al. 1993). By contrast, within the four helical motifs of the hammerhead ribozyme only G^{11} is essential (Joseph et al. 1993).

The sequence requirement within the target is reported as $R^3Y^4N^5G^6H^7Y^8B^9$ (R being A or G; Y being C or U; N being any nucleotide; H being any nucleotide but G; B being any nucleotide but A), the one within the ribozyme as $V^6A^7G^8A^9A^{10}G^{11}Y^{12}$ (V being any nucleotide but U) as suggested by in vitro studies (Joseph et al. 1993). In their in vivo studies, Anderson et al. (1994) have found the substrate sequence $B^4N^5G^6U^7C^8$ to be essential.

6.3 Designing the Appropriate Ribozyme

Although theoretically ribozymes can be designed to target any mRNA, optimal cleavage will only be achieved when several issues are taken into consideration for the ribozyme design (Fig. 2). Since ribozymes other than hammerhead ribozymes share many basic requirements with them (Bratty et al. 1993; Christofferson and Mar 1995), only the criteria for the optimization of hammerhead ribozymes will be discussed here (reviewed by Breaker 1997).

Understanding the biology of a specific system is critical, because the choice of target is crucial and should be a gene known to play a (key) role in cell growth, apoptosis, or differentiation in a specific cell type. Since the triplet GUC is the common cleavage site of many naturally occurring ribozymes, it is often chosen as the site of mRNA digestion in ribozyme design. Although other triplets might also serve as potential cleavage sites, ribozyme activity against them may vary.

The balance between site-specific binding of the ribozyme and the dissociation rate is also of importance: Sufficient turnover rates are achieved when the ribozyme can easily dissociate from the target mRNA after cleavage to repeat the cycle of binding, cleaving, and leaving. This process is supported by short flanking sequences, but these might not provide optimal site-specificity. A ribozyme's capacity for multiple turnovers becomes especially critical when its concentration is considerably lower than that of the target molecules: Abundant mRNAs that are present with about 5000 copies/cell are approximately tenfold more effectively influenced by antisense than by ribozyme molecules (Woolf 1995). However, when the target mRNA has a short half-life or

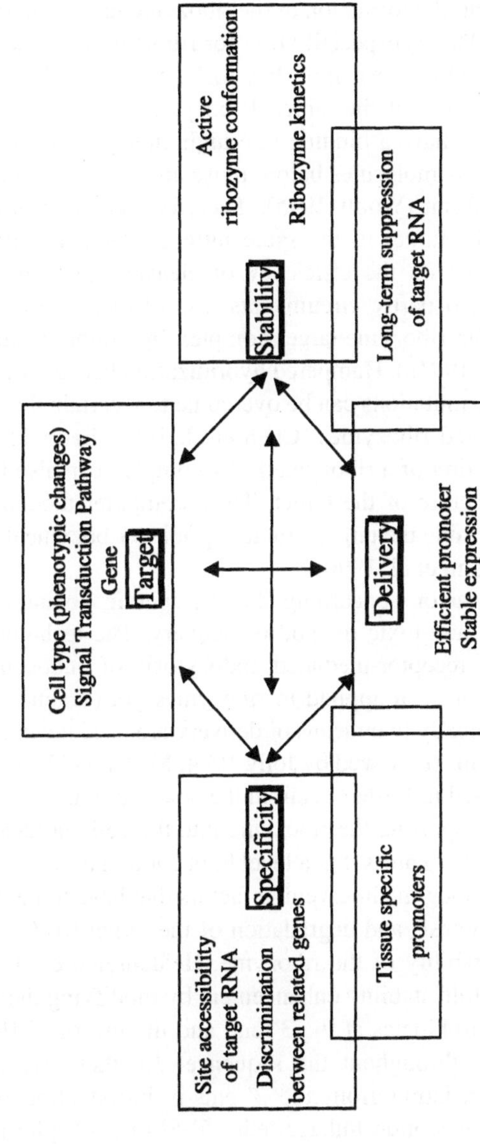

Fig. 2. Designing ribozymes: aspects of consideration (modified from Irie et al. 1997)

is less abundant, the formation of the ribozyme:target complex becomes rate-limiting. This is especially true for ribozymes that act in the cytoplasm, since hybrid formation has to be fast enough to outpace the natural degradation of the target. Ribozymes that are localized in the nucleus need to have a binding rate high enough to cleave substantial amounts of RNA molecules before those are processed and transported into the cytoplasm (Woolf 1995). Therefore, colocalization of the ribozyme and its target in the same intracellular compartment has an immense impact on the efficiency of cleavage (Sullenger and Cech 1993). This also partly circumvents loss of cleavage activity due to digestion of the ribozyme:target complex by double-stranded RNases (Scanlon et al. 1991a). Hampered hybridization because of base-pairing mismatches or mutations can be overcome to a certain degree by multi-unit hammerhead ribozymes (Chen et al. 1992; Leopold et al. 1995). Limited formation of a ribozyme:RNA complex can also be due to the secondary structure of the target. Thus, computer modeling might be useful to determine the target structure prior to ribozyme design (Jaeger et al. 1989; Holm et al. 1996).

A critical factor concerning the therapeutic potential of a given ribozyme is a non-toxic method of delivery. The technology of DNA transfer using receptor-mediated endocytosis of endosomolytic virus proteins has not been applied to ribozymes yet (Wagner et al. 1992). Therefore, basically two means of delivery are used nowadays: nonviral and viral systems (reviewed by Jolly 1994; Miller and Vile 1995). These will be discussed in further detail in the next section.

Besides transporting the ribozyme into the cell, an ideal vector system has to stably express it at a level high enough to reverse the cellular phenotype. Among the intervening factors that have to be overcome are the rate of synthesis and degradation of the target RNA as well as the intracellular stability of the ribozyme. Heidenreich et al. (1994) have reported threefold stability enhancement by modifying the ribozyme by three phosphorothioates at the 3'-end and the use of 2'-fluorocytidine and 2'-uridine throughout the sequence. Another way of inhibiting ribozyme degradation from the 3'-end is introduction of unnatural, inverted internucleotide linkage, e.g., 5'-5' or 3'-3' phosphate diester (Ramalho Ortigao et al. 1992).

In order to express the ribozyme efficiently, an optimal promoter/enhancer system is required. Depending on the target message, factors such as inducible vs constitutive expression, level of expression and tissue specificity must be considered. Constitutive promoters like β-actin (Gunning et al. 1987), cytomegalovirus (CMV; Larsson et al. 1994), or the pol III promoter (Geiduschek and Tocchini-Valentini 1988; Zakharchuk et al. 1995; Thompson et al. 1995) are advantageous if no tissue specificity but high levels of expression are needed. For some diseases, the primary issue of importance might be the design of tissue-specific and/or inducible promoter/enhancer systems (Hart 1996). The tyrosine promoter is an example of the former: specifically expressed in melanoma cells (Vile and Hart 1993) and superior to nonspecific promoters (Ohta et al. 1996a). Vector constructs utilizing heat- or dexamethasone-inducible promoters for ribozyme expression have also been reported (Scanlon et al. 1991b; Zhao and Pick 1993; Ohta et al. 1996b). The recently characterized prostate-specific antigen (PSA) promoter has been shown to exclusively express genes in PSA-producing prostate cancer cells (Pang et al. 1995). In general, the design of a ribozyme against a specific target in a given cell type has to be understood as a unique process. That includes not only optimization of the ribozyme sequence but also of an appropriate expression vector and delivery system.

6.4 Delivery Systems

The apparent importance of gene therapy as an additional form of treatment of cancer or other diseases has led to the development of a variety of techniques for efficient gene delivery and expression (Morgan and Anderson 1993). Those include indirect methods involving ex vivo genetic modification and reimplantation of cells. In terms of ribozyme strategies, direct methods might be the more appropriate. Currently, these can be divided into two major groups: nonviral physical transfection and viral transduction of the target cells.

6.4.1 Viral Vectors

Viral gene transfer is a form of endogenous delivery utilizing the biochemistry of the target cell to express the transferred gene, antisense or ribozyme molecule (Morgan and Anderson 1993; Mulligan 1993).

Thus, the major advantage of this route of delivery is the constitutive intracellular availability of the expressed molecule. Among the viral vectors used in gene therapy studies to date, retroviral and adenoviral systems are the most popular. Wei et al. (1981) described the first retroviral vector, which about a decade later also became the first one used in clinical gene therapy (Anderson et al. 1990). Retroviral vectors are derived from the integrated DNA form of the provirus (Cepko et al. 1984; Miller et al. 1993; Temin 1995). The viral *gag*, *pol*, and *env* genes are essential for virus assembly. They are removed to provide an insertional capacity of up to 8 kb of foreign DNA. Transfections of the vectors into packaging cells expressing *gag*, *pol*, and *env* results in production of viral particles containing the inserted gene. The packaging cells available have been modified to prevent the formation of replication-competent retroviruses (Danos and Mulligan 1988; Dougherty et al. 1989). The fact that retroviral vectors are integrated into the host genome is advantageous when permanent ribozyme expression is essential for alteration of the phenotype (van Beusechem et al. 1992; Dai et al. 1992). However, integration bears the potential safety concern of insertional mutagenesis, although this has not been observed yet. Another major disadvantage of retroviral vectors is the need for high titers to obtain satisfactory transfection efficiency, limiting their usefulness to ex vivo approaches.

Although retroviral vectors are those with which most clinical experience has been accumulated (Jolly 1994), this is changing due to the aforementioned obstacles that can be overcome by adenoviral and other systems (Weitzman et al. 1995). The first adenoviral vectors were reported in 1981 (Thummel et al. 1981; Solnick 1981) and used for the transfer of genes inducing transformation (van Doren et al. 1984; Berkner et al. 1987). Of the 47 known serotypes of adenoviruses, types 2 and 5 (Chroboczek et al. 1992) are the ones most commonly used for vector construction. By replacement of the E1A and E1B genes by the DNA insert to be shuttled, a replication-defective adenovirus is created (Becker et al. 1994). In addition to the safety aspect of E1 removal, this also contributes to nontoxicity of the vectors, avoiding the apparent oncogenic potential of the E1 proteins (DeCaprio et al. 1988; Whyte et al. 1988). These vectors can accommodate up to 7.5 kb of foreign DNA. Theoretically, viral vectors with insertion capacities of 36 kb could be constructed, since only the viral inverted terminal repeats (ITRs) and encapsidation sequences are necessary for replication and packaging. Virus particles are most commonly produced by cotransfecting the construct bearing the foreign DNA with a bacterial helper plasmid carrying the viral flanking sequences into an E1A *trans*-complementing cell line such as 293 (Graham et al. 1977). Thus, titers of replication-defective adenovirus much higher than that of retroviral systems can be produced.

Recently, Heise et al. (1997) have reported a second-generation E1B-attenuated vector that selectively lyses human tumor cells lacking functional p53 protein or expressing human papilloma virus E6 protein (leading to p53 degradation; Werness et al. 1990). Normally, the E1B protein binds to and inactivates wild-type p53 as an essential step for virus replication (Barker and Berk 1987). Consequently, the replication of an adenovirus lacking E1B is critically limited in cells with functional p53, but leads to cytopathic effects comparable to those seen with wild-type adenovirus in p53-deficient cells. This antitumoral effect is selective and has been shown to last up to 6 months in vivo. However, some concerns remain about normal cells not expressing functional p53 and/or not inducing p53 in response to adenoviral infection (MacCallum et al. 1996).

Insertional mutagenesis by adenoviral vectors is not an issue, since they persist as extrachromosomal episomes in the host nucleus (van Doren et al. 1984). A major disadvantage of adenoviral systems, however, is their immunogenic potential (Yang et al. 1994b). Thus, not only repeated injection is limited, but expression of the foreign DNA insert may also be decreased (Yang et al. 1994a). Immunosuppressive conditions might have a positive impact on the clinical use of adenoviral vectors (Engelhardt et al. 1994). In addition, the fact that expression from such vectors is generally transient must be borne in mind during design of a therapy.

A most recent report described the development of an adenoviral/retroviral chimeric vector (Feng et al. 1997), fusing favorable aspects of both systems into a new one. The concept is based on adenoviral vectors that, once they are expressed in the target cell, induce them to function as transient producer cells of retroviral particles. The retroviruses, again, are capable of stably transducing cells in the vicinity. Thus, the adenoviral capacity of efficient gene transfer is combined with the long-term expression of retroviruses.

A third viral vector system lately attracting increasing attention is derived from adeno-associated virus (AAV) type 2, a virus that is not associated with any known human disease (Bartlett et al. 1995). Like retroviruses it provides long-term gene expression by integration into the host genome; although AAV integration is not random, but site-specific into chromosome 19 (Kotin et al. 1990; Samulski 1993). Similar to adenoviruses its infection capacity comprises both dividing and nondividing cells, although the former seem to be somewhat preferred (Russell et al. 1994). Virus particles are usually produced by cotransfection with a helper plasmid providing the genes *rep* and *cap* into adenovirus-infected cells (Nahreini et al. 1993). The first report of high-titer AAV vectors that are independent of helper adenovirus has recently been published (Ferrari et al. 1997). The issues that still have to be addressed before safe and efficient clinical applications can be established are the possibility of cell alteration after wild-type AAV infection (Walz and Schlehofer 1992); the risk of insertional mutagenesis due to leakiness of the site-specificity of integration into chromosome 19 (Muzyczka 1992); and the fact that more than half of all adults show preexisting immunity to AAV (Grossman et al. 1992). Development of other viral systems like those based on the herpes simplex virus (Efstathiou and Minson 1995) or the vaccinia virus (Smith and Moss 1983; Graham et al. 1993) is still in its infancy. Alternative methods of vector delivery also have to be studied in more detail since they might provide as yet undetermined means of anti-tumor vaccination. A candidate for such future methods seems to be "gene painting," i.e. direct application of recombinant adenoviruses onto bare skin (Tang et al. 1997).

6.4.2 Nonviral Systems

The range of nonviral gene delivery methods is broad, including direct injection, particle bombardment with the so-called gene gun, receptor-mediated gene delivery, and lipofection. The first two methods include physical means of overcoming the cell membrane barrier. For particle bombardment, the DNA has to be coated on the surface with gold or tungsten beads that are then accelerated by electric discharges or helium pressure into the target tissue (Yang et al. 1990, 1996; Heiser 1994; Qiu et al. 1996). The only appropriate setting for this method, however, might remain the skin, since for almost all other organs invasive surgery would be required to expose the target tissue well enough (Williams et al. 1991). Direct injection of plasmid DNA seems to be a reasonable way of physical gene delivery, though, since it is simple, safe and nonimmunologic. Although the only tissue known to be susceptible to this method is the striated muscle (Acsadi et al. 1991b), the in vivo results that Acsadi et al. (1991a) have reported for the transfer of the Duchenne muscular dystrophy gene are encouraging. The concept might also hold promise for easy and efficient genetic immunization, as implied by Vahlsing et al. (1994).

More elegant methods involve the generation of DNA:lipid complexes (Lichtenberg 1988; Litzinger and Huang 1992). Immunoliposomes exposing viral protein or antibodies on their surface can be targeted to a specific subset of cells (Leserman et al. 1980; Berinstein et al. 1987; Leonetti and Leserman 1993). Thus, uptake is reported to be enhanced by a factor of 100. Still, release of the DNA from the endosome before its degradation is of concern. One way of facilitating release is by the use of a replication-defective adenovirus that destroys the endosome during acidification (Seth 1994). The use of transferrin-, asialoglycoprotein-, or folate-polylysine-DNA complexes with adenoviral particles has also been reported to result in increased in vitro transfection rates (Curiel et al. 1991; Wagner et al. 1992; Wu et al. 1994; Lasic 1997).

The most important nonviral delivery systems today are based on cationic lipids. One of the first to be developed was N-[1-(2,3-dioleyloxy)propyl]-N,N,N-trimethylammonium chloride (DOTMA; Felgner et al. 1987), now followed by representatives of the next generation (Liu et al. 1997) including (1,2dimyristyloxypropyl-3-dimethyl-hydroxyethyl ammonium bromide (DMRIE; Felgner et al. 1994) and [1,2-dioleoyloxy-3-(trimethylammonio)propane (DOTAP) (Stamatatos et al. 1988; Templeton et al. 1997). Due to the conformational changes that the nucleic acid undergoes in such delivery systems, it is protected against plasma nucleases well enough to allow even systemic delivery.

Several modifications of hammerhead ribozymes have been developed to further enhance stability: Since degradation by RNases involves cleavage of the phosphodiester bond by utilizing the ribose 2'-hydroxyl group, degradation can be inhibited by removal of this group (Taylor et al. 1992) or by introduction of other reactive groups, e.g., 2'-fluoro and/or 2'-amino (Pieken et al. 1991; Heidenreich and Eckstein 1992) or 2'-alkoxy (Goodchild 1992; Paolella et al. 1992). Although the use of cationic lipids is not limited by immunogenicity, integration of the DNA into the genome, carryover of viral proteins or the size of packageable nucleic acid, it has some disadvantages concerning the delivery of ribozymes: The nucleic acid remains trapped in the endosome inhibiting transport into the nucleus. In addition, RNA molecules of less than 500 nucleotides may not be taken up as efficiently as DNA (Christoffersen and Marr 1995). Proof of these concerns is provided by a clinical Phase II trial of melanoma treatment by direct intratumoral injection of a DNA:lipid complex. The transferred plasmid DNA encodes the gene for the class I major histocompatibility complex antigen HLA-B7 (Stopeck et al. 1997; Waddill et al. 1997). Among 36 patients treated so far, 19% showed regression of noninjected nodules, while in 36% local responses of the injected nodules were seen (Hersh et al. 1997).

Still, it cannot be denied that today nonviral methods cannot compete in in vivo settings with viral vector systems. Intensive further study is needed to evaluate the potentials and pitfalls of nonviral systems as tools in gene therapy.

6.5 Targets in Gene Therapy of Malignant and Viral Diseases

Although theoretically ribozymes can be applied to any disease due to altered mRNA expression, it is of critical importance to select target genes and appropriate models that will lead to clinical gene therapy. A suitable target gene would be one known to alter or eradicate cells of the undesired phenotype, thus, the biology and molecular basis of the targeted disease should at least partly be clarified. Since a great number of genes that play a role in carcinogenesis have been identified, many therapeutic ribozyme studies have so far focused on cancer. Among these are anti-cytokine and -oncogene ribozymes as well as ribozymes targeting genes responsible for the development of MDR. Some examples of those studies are given below.

Another field of extensive ribozyme research has been that of viral diseases, primarily the acquired immunodeficiency syndrome related to HIV-1 infection (reviewed by Marschall et al. 1994). However, comparable efforts have also been made using antisense strategies. Thus, HIV studies might become very suitable for comparison of both strategies.

6.5.1 Growth Factors and Cytokines

Signal transduction is initiated by the binding of growth factors to their respective membrane-bound receptor molecules. Alteration of normal signaling pathways is known to lead to development of malignant phenotypes (Seemayer and Cavenee 1989). Therefore, growth factors as molecules primarily involved in transduction initiation are of particular interest as ribozyme targets.

A growth factor polypeptide widely overexpressed by tumor cells is pleiotrophin (PTN). A ribozyme against PTN tested in a PTN-overexpressing melanoma cell line led to inhibition of colony formation. The appropriate control ribozyme did not have such an effect (Czubayko et al. 1994). There is strong evidence that mesothelioma cells are stimulated by autocrine mechanisms involving platelet-derived growth factor-β (PDGF-β) and its specific receptor (Versnel et al. 1988, 1991). In fact, an anti-PDGF-β ribozyme expressed in the target cells via a constitutive vector not only down-regulated PDGF-β mRNA levels but also decreased cell growth. This could not be observed with an inactive control ribozyme (Dorai et al. 1994). Sioud et al. (1992) have used an interesting chemical modification for stabilization of a hammerhead ribozyme cleaving tumor necrosis factor-α (TNFα) mRNA in the promyelocytic cell line HL-60: The 3'-end was protected against nuclease degradation by the T7 transcription terminator. The ribozyme reduced the TNFα mRNA by 90% (vs 40% by a truncated ribozyme) and the protein level to 150 fmol/ml (vs 400 fmol/ml).

6.5.2 Oncogenes

The members of the Ras family are molecules that are central to transduction cascades (Egan and Weinberg 1993). Up to 20% of all malignancies are characterized by mutationally activated Ras (Barbacid 1987). Since the genetic alterations causing these are usually point mutations, the mutated site can serve as a ribozyme target. Hammerhead ribozymes against H-*ras* have been successfully tested in a variety of settings, e.g., in NIH3T3 cells (Koizumi et al. 1992), bladder carcinoma in vitro and in vivo experiments (Kashani-Sabet et al. 1992; Tone et al. 1993; Feng et al. 1995), melanoma cell lines (Ohta et al. 1994, 1996a,b) and in vivo studies (Kashani-Sabet et al. 1994). The latter underline the capability of ribozymes to discriminate between wild-type RNA and its point-mutated counterpart. A second member of the *ras* family, K-*ras*, might become important in the gene therapy of pancreatic cancer, for which no effective therapies exist today. Different delivery systems have been used to target the K-*ras* point mutation in codon 12 found in the vast majority of pancreatic cancers (Grunewald et al. 1989) either with antisense (Aoki et al. 1995) or with ribozyme molecules (Kijima et al. 1996). Again, these studies demonstrate the importance of selecting an appropriate delivery systems. Other examples of solid tumors that can be effectively treated with ribozymes are ovarian and breast cancers overexpressing c-*erb*B-2 (HER-2/neu; Czubayako et al. 1997). Overexpression of the oncogene might not even be a prerequisite for successful ribozyme treatment as implied by in vitro and mice studies with the breast cancer cell line MCF-7 (Suzuki and Anderegg, unpublished data).

Among systemic malignancies, chronic myelogenous and acute lymphoblastic leukemia cells characterized by the *bcr/abl* fusion gene (Philadelphia chromosome) (Rowley 1973) have been the first to be attacked by ribozymes (Wright et al. 1993). Although transcription of the Philadelphia protein was inhibited, normal *bcr* was also cleaved by the ribozyme targeting the *bcr/abl* fusion region. In addition, the secondary structure of the Philadelphia mRNA might be difficult to access by a ribozyme (Pachuk et al. 1993). This problem was circumvented by Gewirtz (1994), who reported decreased expression of *bcr/abl* mRNA antisense treatment against the oncogene c-*myb* — a strategy that might be transferable to ribozyme studies. Another fusion gene, *aml*-1/*eto*, is thought to be relevant in the leukemogenesis leading to acute myelogenous leukemia. Two hammerhead ribozymes have been reported capable of cleaving the chimeric mRNA and inhibiting the leukemic cell growth pattern (Matsushita et al. 1995).

6.5.3 Multidrug Resistance Genes

One of the major drawbacks of cancer chemotherapy is acquired or intrinsic drug resistance, often in the form of MDR against a variety of compounds (Endicott and Ling 1989; Gottesman and Pastan 1993). Probably in the majority of MDR cases, the reason is overexpression of the gene for P-glycoprotein, defined as the *mdr-1* (Roninson et al. 1986). Anti-*mdr-1* ribozymes have been successfully applied to reverse the MDR phenotype through down-regulation of *mdr-1* mRNA and P-glycoprotein (Scanlon et al. 1994; Kobayashi et al. 1994; Bouffard et al. 1996; Ohkawa et al. 1996). These studies underline the fact that prediction of the most effective target site is not possible and that selection of an optimized ribozyme requires careful evaluation in the cell of interest.

Besides direct targeting of the *mdr*-1 gene, cleavage of other genes playing a role in MDR development has also been demonstrated to be effective. The promoter/enhancer element of *mdr*-1 contains a binding site for the activator protein-1 (AP-1), a heterodimeric complex of proteins of the Fos and Jun families. Therefore, the activation of *fos* and/or *jun* is also thought to play a role in MDR (Ueda et al. 1987). In fact, an anti-c-*fos* ribozyme not only down-regulated the expression of the target gene, but also had an impact on the level of *mdr*-1 mRNA which was even stronger than that of an anti-*mdr*-1 ribozyme (Scanlon et al. 1991a, 1994; Funato et al. 1992). Thus, targeting of pivotal genes throughout the signal transduction pathway might result in successful reversal of the MDR phenotype. Some cells with MDR phenotype do not overexpress P-glycoprotein, but rather the MDR-associated protein MRP (Cole et al. 1992) or the lung resistance-related protein LRP (Scheper et al. 1993). These might also serve as adequate targets for ribozyme-mediated restoration of drug sensitivity of cancer cells.

6.5.4 Human Immunodeficiency Virus-1

Some of the most interesting HIV-1 studies in terms of design and expression of the most effective ribozymes as well as interpretation of the results are described in the following. The so-called catalytic antisense RNA used by Homann et al. (1993) consists of a long RNA sequence (about 400 nucleotides) encompassing an anti-*gag* ribozyme.

Although an inactive control ribozyme also reduced HIV-1 replication in vitro, probably due to an antisense effect of the long flanking sequences, the catalytic antisense RNA was four- to sevenfold more efficient. A direct comparison between an antisense ribonucleotide and a ribozyme, both targeting the HIV-1 *tat* sequence, showed even superior inhibition of HIV-1 particle production by the antisense oligonucleotide (Lo et al. 1992).

Some of the studies seem to be of special interest for clinical applications of anti-HIV-1 ribozymes. The gene for the envelope protein *env* has been targeted with a plasmid containing up to nine ribozymes in a stretch. It was significantly more efficient than a monoribozyme (Chen et al. 1992). The results of in vitro experiments have not only resulted in drastic inhibition of HIV replication, but also in excessive reduction of syncytia formation. However, another study has shown that a mixture of five monoribozymes was more active than the corresponding pentaribozyme (Bertrand et al. 1994). A hairpin ribozyme directed against the HIV-1 leader sequence (Ojwang et al. 1992) seems to be especially promising as a potential tool for clinical gene therapy, since it is capable of cleaving diverse HIV strains (Yu et al. 1993). Still, it has to be noted that some ribozymes have been shown to be inefficient tools in vitro but effective inhibitors of virus replication in human HIV-1 infected T cells in vivo (Crisell et al. 1993). Thus, in vitro results have to be interpreted very carefully, since the performance of a particular ribozyme in an in vitro setting cannot necessarily be transferred to the in vivo situation (Woolf 1995).

6.6 Summary and Outlook

Ribozymes are effective tools for the modulation of specific gene expression due to their site-specific cleavage activity and their potential to be targeted against virtually any mRNA linked to a disease. Malignant and viral diseases are obvious targets for ribozyme strategies. Various studies have already demonstrated their value concerning reversal of the abnormal phenotype. However, data derived from in vitro experiments does not always correlate with in vivo data. Therefore, it is highly desirable to extend the work done on ribozymes so far to more and clinically relevant in vivo studies. Meanwhile, further efforts to substantiate the understanding of the molecular processes, alterations, and pathways leading to deregulation of cell growth and differentiation, i.e., underlying cell transformation, will be required. Thus, new targets for ribozyme strategies will be identified. Another important issue to be addressed in the near future is the further development of efficient and appropriate vector systems. In addition, for the large number of diseases potentially amenable to ribozyme therapy, it would be advantageous to find suitable ways of efficient and specific systemic delivery. Especially to this end, better means of ribozyme resistance against degradation have to be found. Still, it seems already clear that ribozymes are an important addition to the armory for gene expression interference. Moreover, they might serve as another tool in defining the role specific oncogenes play in specific types of tumors.

Acknowledgements. We are grateful to Dr. David Henderson for critically reading the manuscript.

References

Acsadi G, Dickson G, Love DR, Jani A, Walsh FS, Gurusignhe A, Wolff JA, Davies KE (1991a) Human dystrophin expression in mdx mice after intracellular injection of DNA constructs. Nature 352:815–818

Acsadi G, Jiao S, Jani A, Duke D, Williams P, Chong W, Wolff JA (1991b) Direct gene transfer and expression into heart in vivo. New Biol 3:71–81

Anderson P, Monforte J, Tritz R, Nesbitt S, Hearst J, Hampel A (1994) Mutagenesis of the hairpin ribozyme. Nucleic Acids Res 22:1096–1100

Anderson WF, Blaese RM, Culver K (1990) The ADA human gene therapy protocol. Hum Gene Ther 1:331–362

Aoki K, Yoshida T, Sugimura T, Terada M (1995) Liposome-mediated in vivo gene transfer of antisense K-ras construct inhibits pancreatic tumor dissemination in the murine peritoneal cavity. Cancer Res 55:3810–3816

Barbacid M (1987) ras genes. Annu Rev Biochem 56:779–827

Barker DD, Berk AJ (1987) Adenovirus proteins from both E1B reading frames are required for transformation of rodent cells by viral infection and DNA transfection. Virology 156:107–121

Bartlett JS, Quattrocchoni KB, Samulski RJ (1995) The development of adeno-associated virus as a vector for cancer gene therapy. In: Sobol RE, Scanlon KJ (eds) The internet book of gene therapy. Appleton and Lange, Stamford, pp 27–39

Becker TC, Noel RJ, Coats WS, Gomez-Foix AM, Alam T, Gerard RD, Newgard CB (1994) Use of recombinant adenovirus for metabolic engineering of mammalian cells. Methods Cell Biol 43:161–189

Beigelman L, McSwiggen JA, Draper KG, Gonzalez C, Jensen K, Karpeisky AM, Modak AS, Matulic-Adamic J, DiRenzo AB, Haeberli P, Sweedler D, Tracz D, Grimm S, Wincott FE, Thackray VG, Usman N (1995) Chemical modification of hammerhead ribozyme. J Biol Chem 270:25702–25708

Bennett MJ, Cullimore JV (1992) Selective cleavage of closely related mRNAs by synthetic ribozymes. Nucleic Acids Res 20: 831–837

Berinstein N, Matthay KK, Papaphadjopoulos D, Levy R, Sikic BI (1987) Antibody-directed targeting of liposomes to human cell lines: role of binding and internalization in growth inhibition. Cancer Res 47:5954–5959

Berkner KL, Schaffhausen BS, Roberts TM, Sharp PA (1987) Abundant expression of polyomavirus middle T antigen and dihydrofolate reductase in an adenovirus recombinant. J Virol 61:1213–1220

Bertrand E, Pictet R, Grange T (1994) Can hammerhead ribozymes be efficient tools to inactivate gene function? Nucleic Acids Res 22:293–300

Berzal-Herranz A, Joseph S, Chowrira BM, Butcher SE, Burke JM (1993) Essential nucleotide sequences and secondary structure elements of the hairpin ribozyme. EMBO J 12:2564–2567

Boiziau C, Kufurst R, Cazenave C, Roig V, Thuong NT, Toulme J-J (1991) Inhibition of translation by antisense oligonucleotides via an Rnase-H independent mechanism. Nucleic Acids Res 19:1113–1119

Bouffard DY, Ohkawa T, Kijima H, Irie A, Suzuki T, Curcio LD, Holm PS, Sassani A, Scanlon KJ (1996) Oligonucleotide modulation of multidrug resistance. Eur J Cancer 32A:1010–1018

Branch AD, Robertson HD (1991) Efficient trans cleavage and common structural motif for the ribozymes of the human hepatitis δ agent. Proc Natl Acad Sci USA 88:10163–10167

Bratty J, Chartrand P, Ferbeyre G, Cedergren R (1993) The hammerhead RNA domain, a model ribozyme. Biochem Biophys Acta 1216:345–359

Breaker RR (1997) In vitro selection of catalytic polynucleotides. Chem Rev 97:371–390

Buzayan JM, Hampel A, Brüning G (1986) Nucleotide sequence and newly formed phosphodiester bond of spontaneously limaged satellite tobacco ringspot virus RNA. Nucleic Acids Res 14:9729–9743

Buzayan JM, Feldstein PA, Bruening G, Eckstein F (1988) RNA mediated formation of a phospho diester bond. Biochem Biophys Res Commun 156:340–347

Cameron FH, Jennings PA (1989) Specific gene suppression by engineered ribozymes in monkey cells. Proc Natl Acad Sci USA 86:9139–9143

Cech TR, Zaug AJ, Grabowski PJ (1981) In vitro splicing of the ribosomal RNA precursor of Tetrahymena: involvement of a guanosine nucleotide in the excision of the intervening sequence. Cell 27:1053–1062

Cepko CL, Roberts BE, Mulligan RC (1984) Construction and applications of a highly transmissible murine retrovirus shuttle vector. Cell 37:1053–1062

Chartrand P, Harvey SC, Febeyre G, Usman N, Cedergreen R (1995) An oligonucleotide that supports catalytic activity in the hammerhead ribozyme domain. Nucleic Acids Res 23:4091–4096

Chen C-J, Banerjea ACA, Hrmision GG, Haglund K, Schubert M (1992) Multitarget-ribozymes directed to cleave at up to nin highly conserved HIV-1 env RNA regions inhibit HIV-1 replication – potential effictiveness against most presently sequenced HIV-1 isolates. Nucleic Acids Res 20:4581–4589

Chowrira BM, Berzal-Herranz A, Burke JM (1991) Novel guanosin requirement for catalysis by the hairpin ribozyme. Nature 354:320–322

Chowrira BM, Berzal-Herranz A, Keller CF, Burke JM (1993) Four ribose 2'-hydroxyl groups essential for the catalytic function of the hairpin ribozyme. J Biol Chem 268:19458–19462

Christoffersen RE, Marr JJ (1995) Ribozymes as human therapeutic agents. J Med Chem 38:2023–2037

Chroboczek J, Bieber F, Jacrot B (1992) The sequence of the genome of adenovirus type 5 and its comparison with the genome of adenovirus type 2. Virology 186:280–285

Cole SPC, Bhardwaj G, Gerlach JH, Mackie JE, Grant CE, Almquist KC, Stewart AJ, Kurz EU, Duncan AMV, Deeley RG (1992) Overexpression of a transporter gene in multidrug-resistant human lung cancer cell line. Science 258:1650–1654

Crisell P, Thompson S, James W (1993) Inhibition of HIV-1 replication by ribozymes that show poor activity in vitro. Nucleic Acids Res 21:5251–5255

Curcio LD, Bouffard DY, Scanlon KJ (1997) Oligonucleotides as modulators of cancer gene expression. Pharmacol Ther 74:317–332

Curiel DT, Agarwal S, Wagner E, Cotten M (1991) Adenovirus enhancement of transferrin-polylysine-mediated gene delivery. Proc Natl Acad Sci USA 88:8850–8854

Czubayko F, Riegel AT, Wellstein A (1994) Ribozyme-targeting elucidates a direct role of pleiotrophin in tumor growth. J Biol Chem 269:21358–21363

Czubayako F, Downing SG, Hsieh SS, Goldstein DJ, Lu PY, Trapnell BC, Wellstein A (1997) Adenovirus-mediated transduction of ribozymes abrogates HER-2/neu and pleitrophin expression and inhibits tumor cell proliferation. Gene Ther 4:943–949

Dai Y, Roman M, Naviaux RK, Verma IM (1992) Gene therapy via primary myoblasts: long-term expression of factor IX protein following transplantation in vivo. Proc Natl Acad Sci USA 89:10892–10895

Danos O, Mulligan RC (1988) Safe and efficient generation of recombinant retroviruses with amphotropic and ecotropic host ranges. Proc Natl Acad Sci USA 85:6460–6464

DeCaprio JA, Ludlow JW, Figge J, Stein JY, Huang CM, Lee WH, Marsilio E, Paucha E, Livingston DM (1988) SV40 large tumor antigen forms a specific complex with the product of the retinoblastoma susceptibility gene. Cell 54:275–238

Demidov V, Frank-Kamentskii MD, Egholm M, Buchardt O, Nielsen PE (1993) Sequence selective double strand DNA cleavage by peptide nucleic acid (PNA) targeting using nuclease S1. Nucleic Acids Res 21:2103–2107

Dorai T, Kobayashi H, Holland JF, Ohnuma T (1994) Modulation of platelet-derived growth factor-beta mRNA expression and cell growth in a human mesothelioma cell line by a hammerhead ribozyme. Mol Pharmacol 96:437–444

Dougherty JP, Wisnewski R, Yang S, Yang SL, Rhoode BW, Temin HM (1989) New retrovirus helper cells with almost no nucleotide homology to retrovirus vectors. J Virol 63:3209–3212

Efstathiou S, Minson AC (1995) Herpes virus based vectors. Br Med Bull 51:45–55

Egan SE, Weinberg RA (1993) The pathway to signal achievement. Nature 365:781–783

Endicott JA, Ling V (1989) The biochemistry of P-glycoprotein mediated multidrug resistance. Annu Rev Biochem 58:131–171

Engelhardt JF, Ye X, Doranz B, Wilson JM (1994) Ablation of E2 A in recombinant adenovirus improves transgene persistence and decreases inflammatory response in mouse liver. Proc Natl Acad Sci USA 91:6196–6200

Fearon ER, Vogelstein B (1990) A genetic model for colorectal tumorigenesis. Cell 61:759–767

Feldstein PA, Buzayan JM, Bruening G (1989) Two sequences participating in the autolytic processing of satellite tobacco ringspot virus complementary RNA. Gene 82:53–61

Felgner JH, Kumar R, Sridhar CM, Wheeler CJ, Tsai YJ, Border R, Ramsey P, Martin M, Felgner PL (1994) Enhanced gene delivery and mechanism studies with a novel series of cationic lipid formulations. J Biol Chem 269:2550–2561

Felgner PL, Gadek TR, Holm M, Roman R, Chan HW, Wenz M, Northrop JP, Ringold GM (1987) Lipofection: a highly efficient lipid-mediated DNA-transfection procedure. Proc Natl Acad Sci USA 84:7413–7417

Felsenfeld G, Davies DR, Rich A (1957) Formation of a three-stranded polynucleotide molecule. J Am Chem Soc 79:2023–2024

Feng M, Cabrera G, Deshane J, Scanlon KJ, Curiel DT (1995) Neoplastic reversion accomplished by high efficiency adenoviral mediated delivery of anti-ras ribozymes. Cancer Res 55:2024–2028

Feng M, Jackson WH Jr, Goldman CK, Rancourt C, Wang M, Dusing SK, Siegal G, Curiel DT (1997) Stable in vivo gene transduction via a novel adenoviral/retroviral chimeric vector. Nat Biotech 15:866–870

Ferrari FK, Xiao X, McCarty D, Samulski RJ (1997) New developments in the generation of Ad-free, high-titer rAAV gene therapy vectors. Nat Med 3:1295–1297

Forster AC, Symons RH (1987) Self-cleavage of plus and minus RNAs of a virusoid and structural model for the active sites. Cell 49:211–220

Foulds L (1958) The natural history of cancer. J Chronic Dis 8:2–37

Funato T, Yoshida E, Jiao L, Tone T, Kashani-Sabet M, Scanlon K (1992) The utility of the anti-fos ribozyme in reversing cis-platin resistance in human carcinomas. Adv Enzyme Regul 32:195–209

Geiduschek EP, Tocchini-Valentini GP (1988) Transcription by RNA polymerase III. Annu Rev Biochem 57:873–914

Gerlach WL, Llewellyn D, Haseloff J (1987) Construction of a plant disease resistance gene form the satellite RNA of tobacco ring spot virus. Nature 328:802–805

Gewirtz AM (1994) Treatment of chronic myelogenous leukemia (CML) with c-myb antisense oligodeoxynucleotides. Bone Marrow Transplant 14 [Suppl 3]:57–61

Goodchild J (1992) Enhancement of ribozyme catalytic activity by a contiguous oligodeoxynucleotide (facilitator) and by 2'-O-methylation. Nucleic Acids Res 20:4607–4612

Gottesman MM, Pastan I (1993) Biochemistry of multidrug resistance mediated by the multidrug transporter. Annu Rev Biochem 62:385–427

Graham BS, Matthews TJ, Belshe RB, Clements ML, Dolin R, Wright PF, Gorse GJ, Schwarts DH, Keefer MC, Bologhesi DP (1993) Augmentation of human immunodeficiency virus type 1 neutralizing antibody by priming with gp160 recombinant vaccinia and booting with rgp160 in vaccinia-naive adults. J Infect Dis 167:533–537

Graham FL, Smiley J, Russel WC, Nairn R (1977) Characteristics of a human cell line transformed by DNA from human adenovirus 5. J Gen Virol 36:59–74

Grossman Z, Mendelson E, Brok-Simoni F, Milegvir F, Leifner Y, Rechavi G, Ramot B (1992) Detection of adeno-associated virus type 2 in human peripheral blood cells. J Gen Virol 73:961–966

Grunewald K, Lyons J, Froehlich A, Feichtinger H, Weger RA, Schwab G, Janssen JW, Bartram CR (1989) High frequency of Ki-ras codon 12 mutations in pancreatic adenocarcinomas. Int J Cancer 43:1037–1041

Guerrier-Takada C, Gardiner K, Marsh T, Pace N, Altman S (1983) The RNA moiety of ribonuclease P is the catalytic subunit of the enzyme. Cell 35:849–857

Gunning P, Leavitt J, Muscat G, Ng S, Kedes L (1987) A human beta-actin expression vector systems directs high-level accumulation of antisense transcripts. Proc Natl Acad Sci USA 84:4831–4835

Hampel A, Tritz R (1989) RNA catalytic properties of the minimum (-)sTRSV sequence. Biochemistry 28:4929–4933

Hampel A, Tritz R, Hicks M, Cruz P (1990) "Hairpin" catalytic RNA model: evidence for helices and sequence requirement for substrate RNA. Nucleic Acids Res 18:299–304

Hart IR (1996) Tissue specific promoters in targeting systemically delivered gene therapy. Semin Oncol 23:54–158

Haseloff J, Gerlach WL (1988) Simple RNA enzymes with new and highly specific endoribonuclease activities. Nature 334:585–591

Heidenreich O, Benseler F, Fahrenholy A, Eckstein F (1994) High activity and stability of hammerhead ribozymes containing 2'-modified pyrimidine nucleosides and phosphorothioates. J Biol Chem 269:2131–2138

Heidenreich O, Eckstein F (1992) Hammerhead ribozyme-mediated cleavage of the long terminal repeat RNA of human immunodeficiency virus type 1. J Biol Chem 267:1904–1909

Heise C, Sampson-Johannes A, Williams A, McCormick F, von Hoff DD, Kirn DH (1997) ONYX-015, an E1B-attenuated adenovirus, causes turmor-specific cytolysis and antitumoral efficacy that can be augmented by standard chemotherapeutic agents. Nat Med 3:639–645

Heiser WC (1994) Gene transfer into mammalian cells by particle bombardment. Anal Biochem 217:185–196

Hendry P, McCall MJ, Santiago FS, Jennings PA (1992) A ribozyme with DNA in the hybridizing arms displays enhanced cleavage ability. Nucleic Acids Res 20:5737–5741

Hendry P, McCall MJ, Santiago FS, Jennings PA (1995) In vitro activity of minimized hammerhead ribozyme. Nucleic Acids Res 23:3922–3927

Herschlag D (1991) Implications of ribozyme kinetics for targeting the cleavage of specific RNA molecules in vivo: more isn't always better. Proc Natl Acad Sci USA 88:6921–6925

Hersh EM, Stopeck AT, Silver HKB, Chang AE, Doroshow JH, Rubin J, Rhinehart J, Stahl SM, Schreiber AB (1997) Phase II study of intratumoral injection of HLA-B7/β2 M plasmid DNA-cationic lipid complex (Allovectin-7) therapy for metastatic malignant melanoma (MMM). Cancer Gene Ther 4 [Suppl]:48–49 (abstract)

Hertel KJ, Pardi A, Uhlenbeck OC, Koizumi M, Ohtsuka E, Uesugi S, Cedergreen R, Eckstein F, Gerlach WL, Hodgson R, Symons RH (1992) Numbering system for the hammerhead. Nucleic Acids Res 20:3252

Hertel KJ, Herschlag D, Uhlenbeck OC (1994) A kinetic and thermodynamic framework for the hammerhead ribozyme reaction. Biochemistry 33:3374–3385

Holm PS, Dietel M, Krupp G (1996) Similar cleavage efficiencies of an oligoribonucleotide substrate and an mdr1 mRNA segment by a hammerhead ribozyme. Gene 167:221–225

Homann M, Tzortzakaki S, Rittner K, Sczkiel G, Tabler M (1993) Incorporation of the catalytic domain of a hammerhead ribozyme into antisense RNA enhances its inhibitory effect on the replication of human immunodeficiency virus type I. Nucleic Acids Res 21:2809–2814

Hutchins CJ, Rathjen PD, Forster AC, Symons RH (1986) Self-cleavage of plus and minus RNA transcripts of avocado sun blotch viroid. Nucleic Acids Res 14:3627–3640

Irie A, Kijima H, Ohkawa T, Bouffard DY, Suzuki T, Curcio LD, Holm PS, Sassani A, Scanlon KJ (1997) Anti-oncogene ribozymes for cancer gene therapy. Adv Pharmacol 40:207–257

Jaeger JA, Turner DH, Zuker M (1989) Improved predictions of secondary structures for RNA. Proc Natl Acad Sci USA 86:7706–7710

Jolly D (1994) Viral vector systems for gene therapy. Cancer Gene Ther 1:51–64

Joseph S, Berzal-Herranz A, Chowrira BM, Butcher SE, Burke JM (1993) Substrate selection rules for the hairpin ribozyme determined by in vitro selection, mutation, and analysis of mismatched substrates. Genes Dev 7:130–138

Kashani-Sabet M, Funato T, Tone T, Jiao L, Wang W, Yoshida E, Kashfian BI, Shitara T, Wu AM, Moreno JG, Traweek ST, Ahlering TE, Scanlon KJ (1992) Reversal of the malignant phenotype by an anti-ras ribozyme. Antisense Res Dev 2:3–15

Kashani-Sabet M, Funato T, Florenes VA, Fodstad O, Scanlon KJ (1994) Suppression of the neoplastic phenotype in vivo by an anti-ras ribozyme. Cancer Res 54:900–902

Kijima H, Bouffard DY, Scanlon KJ (1996) Ribozyme-mediated reversal of human pancreatic carcinoma phenotype. In: Ikehara S (ed) Proceedings of international symposium on bone marrow transplantation. Springer, Berlin Heidelberg New York, pp 153–163

Kobayashi H, Dorai T, Holland JF, Ohnuma T (1994) Reversal of drug sensitivity in multidrug-resistant tumor cells by MDR-1 (PGY-1) ribozyme. Cancer Res 54:1271–1275

Koizumi M, Hayase Y, Iwai S, Kamiya H, Inoue H, Ohtsuka E (1992) Design of RNA enzymes distinguishing a single base mutation in RNA. Nucleic Acids Res 17:7059–7071

Kotin RM, Siniscalco M, Samulski RJ, Zhu XD, Hunter L, Laughlin CA, McLaughlin S, Muzyczka N, Rocchi M, Berns KI (1990) Site-specific integration by adeno-associated viurs. Proc Natl Acad Sci USA 87:2211–2215

Kruger K, Grabowski PJ, Zaug AJ, Sands J, Gottschling DE, Cech TR (1982) Self-splicing RNA: autoexcision and autocyclization of the ribosomal RNA intervening sequence of Tetrahymena. Cell 31:147–157

Kumar PKR, Ellington AD (1995) Artificial evolution and natural ribozymes. FASEB J 9:1183–1195

Larsson S, Hotchkiss G, Andang M, Nyholm T, Inzunza J, Jansoon I, Ahrlund-Richter L (1994) Reduced β-microglobulin mRNA levels in transgenic mice expressing a designed hammerhead ribozyme. Nucleic Acids Res 22:2242–2248

Lasic DD (1997) Recent developments in medical applications of liposomes: sterically stabilized liposomes in cancer therapy and gene delivery in vivo. J Contr Release 48:203–222

Leonetti JP, Leserman LD (1993) Targeted delivery of oligonucleotides. In: Cook ST, Lebleu B (eds) Antisense research and applications. CRC Press, Boca Raton, pp 493–504

Leopold LH, Shore SK, Newkirk TA, Reddy RMV, Reddy EP (1995) Multi-unit ribozyme mediated cleavage of bcr-abl mRNA in myeloid leukemias. Blood 85:2162–2170

Leserman LD, Weinstein JN, Blumenthal R, Terry WD (1980) Receptor-mediated endocytosis of antibody-opsonized liposomes by tumor cells. Proc Natl Acad Sci USA 77:4089–4093

Lichtenberg D (1988) Liposomes: preparation, characterization, and preservation. Method Biochem Anal 33:337–468

Litzinger DC, Huang L (1992) Phosphatidylethanolamine liposomes: drug delivery, gene transfer and immunodiagnostic applications. Biochim Biophys Acta 1113:201–227

Liu Y, Mounkes LC, Liggitt HD, Brown CS, Solodin I, Heath TD, Debs RJ (1997) Factors influencing the efficiency of cationic liposome-mediated intravenous gene delivery. Nat Biotechnol 15:167–173

Lo KM, Biasolo MA, Dehni G, Palu G, Haseltine WA (1992) Inhibition of replication of HIV-1 by retroviral vectors expression tat-antisense and anti-tat ribozyme RNA. Virology 190:176–183

MacCallum DE, Hupp TR, Midgley CA, Stuart D, Campbell SJ, Harper A, Walsh FS, Wright EG, Balmain A, Lane DP, Hall PA (1996) The p53 response to ionizing radiation in adult and developing tissues. Oncogene 13:2575–2587

Marschall P, Thomson JB, Eckstein F (1994) Inhibition of gene expression with ribozymes. Cell Mol Neurobiol 14:523–538

Matsushita H, Kobayashi H, Mori S, Kizaki M, Ikeda Y (1995). Ribozymes cleave the AML1/MTG8 fusion transcript and inhibit proliferation of leukemic cells with t(8;21). Biochem Biophys Res Commun 215:431–437

McCall MJ, Hendry P, Jennings PA (1992) Minimal sequence requirements for ribozyme activity. Proc Natl Acad Sci USA 89:5710–5714

McKay DB (1996) Structure and function of the hammerhead ribozyme: an unfinished story. RNA 2:395–403

Melton DA (1985) Injected antisense RNAs specifically block messenger RNA translation in vivo. Proc Natl Acad Sci USA 82:144–148

Miller AD, Miller DG, Garcia JV, Lynch CM (1993) Use of retroviral vectors for gene transfer and expression. In: Wu R (ed) Methods in enzymogoly, vol 217. Academic, San Diego, pp 581–599

Miller N, Vile R (1995) Targeted vectors for gene therapy. Blood 76:271–599

Morgan AR, Wells RD (1968) Specificity of the three-stranded complex formation between double-stranded DNA and single-stranded RNA containing repeating nucleotide sequences. J Mol Biol 37:63–80

Morgan RA, Anderson WF (1993) Human gene therapy. Annu Rev Biochem 62:191–217

Mulligan RC (1993) The basic science of gene therapy. Science 260:926–932

Muzyczka N (1992) Use of adeno-associated virus as a general transduction vector for mammalian cells. In: Capron A, Compans RW, Cooper M et al (eds) Current topic in microbiology and immunology, vol 158. Springer, Berlin Heidelberg New York, pp 97–129

Nahreini P, Woody MJ, Zhou SZ, Srivastava A (1993) Versatile adeno-associated virus 2-based vectors for constructing recombinant virions. Gene 124:257–262

Nakamaye KL, Eckstein F (1994) AUA-cleaving hammerhead ribozymes: attempted selection for improved cleavage. Biochemistry 33:1271–1277

Ohkawa T, Koguma T, Okhada T, Taira K (1995) Ribozyme: from mechanistic studies to applications in vivo. J Biochem 118:251–258

Ohkawa T, Kijima H, Irie A, Horng G, Kaminski A, Tsai J, Kashfian BI, Scanlon KJ (1996) Oligonucleotide modulation of multidrug resistance gene expression. In: Gupta S, Tsuruo T (eds) Multidrug resistance in cancer cells: cellular, biochemical, molecular and biological aspects. Wiley, New York, pp 413–433

Ohta Y, Tone T, Shitara T, Funato T, Jiao L, Kashfian BI, Yoshida E, Horng M, Tsai P, Lauterbach K, Kashani-Sabet M, Florenes VA, Fodstad OY, Scanlon KJ (1994) H-ras ribozyme mediated alteration of the human melanoma phenotype. Ann NY Acad Sci 716:242–253

Ohta Y, Kijima H, Kashani-Sabet M, Scanlon KJ (1996a) Tissue-specific expression of an anti-ras ribozyme inhibits proliferation of human malignant melanoma cells. Nucleic Acids Res 24:938–942

Ohta Y, Kijima H, Kashani-Sabet M, Scanlon KJ (1996b) Suppresion of the malignant phenotype of melanoma cells by anti-oncogene ribozymes. J Invest Dermatol 106:275–280

Ojwang JO, Hampel A, Looney DJ, Wong SF, Rappaport J (1992) Inhibition of human immunodeficiency virus type 1 expression by a hairpin ribozyme. Proc Natl Acad Sci USA 89:10802–10806

Pachuk CJ, Yoon K, Moelling K, Coney LR (1993) Selective cleavage of bcr-abl chimeric RNAs by a ribozyme targeted to noncontiguous sequences. Nucleic Acids Res 22:301–307

Pang S, Taneja S, Dardashi K, Cohan P, Kaboo R, Sokoloff M, Tso C, DeKernion JB, Belldegrun AS (1995) Prostate tissue specificity of the prostate-specific antigen promoter isolated from a patient with prostate cancer. Human Gene Ther 6:1417–1426

Paolella G, Sproat B, Lamond AI (1992) Nuclease resistant ribozymes with catalytic activity. EMBO J 11:1913–1919

Perotta AT, Been MD (1992) Cleavage of oligoribonucleotides by a ribozyme derived from the hepatitis δ virus RNA sequence. Biochemistry 31:16–21

Perriman R, Delves A, Gerlach WL (1992) Extended target-site specificity for a hammerhead ribozyme. Gene 113:157–163

Pieken WWA, Olsen DB, Benseler F, Aurup H, Eckstein F (1991) Kinetic characterization of ribonuclease resistant 2'-modified hammerhead ribozymes. Science 253:314–317

Pley HW, Flaherty KM, McKay DB (1994) Three-dimensional structure of a hammerhead ribozyme. Nature 372:68–74

Pyle A (1993) Ribozyme: a distinct class of metalloenzymes. Science 261:709–714

Qiu P, Ziegelhoffer P, Sun J, Yang NS (1996) Gene gun delivery of mRNA in situ results in efficient transgene expression and genetic immunization. Gene Ther 3:262–268

Ramalho Ortigao JF, Rosch H, Selter H, Frohlich A, Lorenz A, Montenath M, Seliger H (1992) Antisense effect of oligodeoxynucleotides with inverted terminal internucleotidic linkages: a minimal modification protecting against nucleolytic degradation. Antisense Res Dev 2:129–146

Roninson IB, Chin JE, Choi K, Gros P, Housman DE, Fojo A, Gottesman MM, Pastan I (1986) Isolation of human mdr DNA sequences amplified in multidrug resistant KB carcinoma cells. Proc Natl Acad Sci USA 83:4538–4542

Rowley JD (1973) A new consistent chromosomal abnormality in chronic myelogenous leukaemia identified by quinacrine fluorescence and Giemsa staining. Nature 243:290–293

Ruffner DE, Stormo GD, Uhlenbeck OC (1990) Sequence requirements of the hammerhead RNA self-cleavage reaction. Biochemistry 29:10695–10702

Russell DW, Miller AD, Alexander IE (1994) Adeno-associated virus vectors preferrentially transduce cells in S phase. Proc Natl Acad Sci USA 91:8915–8919

Samulski RJ (1993) Adeno-associated virus: integration at a specific chromosomal locus. Curr Opin Biotech 3:74–80

Scanlon KJ, Jiao L, Funato T, Wang W, Tone T, Rossi JJ, Kashani-Sabet M (1991a) Ribozyme-mediated cleavage of c-fos mRNA reduces gene expression of DNA synthesis enzymes and metallothionein. Proc Natl Acad Sci USA 88:10591–10595

Scanlon KJ, Kashani-Sabet M, Tone T, Funato T (1991b) Cisplatin resistance in human cancers. Pharmacol Ther 52:385–406

Scanlon KJ, Ishida H, Kashani-Sabet M (1994) Ribozyme-mediated reversal of the multidrug-resistant phenotype. Proc Natl Acad Sci USA 91:11123–11127

Scheper RJ, Broxterman HJ, Scheffer GL, Kaaijk P, Dalton WS, van Heijningen THM, van Kalken CK, Slovak ML, de Vries EGE, van der Valk P, Meijer CJLM, Pinedo HM (1993) Overexpression of a Mr 110,000 vesicular protein in non-P-glycoprotein-mediated multidrug resistance. Cancer Res 53:1475–1479

Scott WG, Finch JT, Klug A (1995) The crystal structure of an all-RNA hammerhead ribozyme: a proposed mechanism for RNA catalytic cleavage. Cell 81:991–1002

Seemayer TA, Cavenee WK (1989) Biology of diseases: molecular mechanisms of oncogenesis. Lab Invest 60:585–599

Seth P (1994) Adenovirus-dependent release of choline from plasma membrane vesicles at an acidic pH is mediated by penton base protein. J Virol 68:1204–1206

Shimayama T, Wishikawa F, Nishidawa S, Taira K (1993) Nuclease-resistant chimeric ribozymes containing deoxyribonucleotides and phosphorothioate linkages. Nucleic Acids Res 21:2605–2611

Sigurdsson STH, Tuschl T, Eckstein F (1995) Proving RNA tertiary structure: interhelical crosslinking of the hammerhead ribozyme. RNA 1:575–583

Sioud M, Natvig JB, Forre O (1992) Preformed ribozyme destroys tumor necrosis factor mRNA in human cells. J Mol Biol 223:831–835

Smith GL, Moss B (1983) Infectious poxvirus vectors have a capacity for at least 25,000 base pairs of foreign DNA. Gene 25:21–28

Solnick D (1981) Construction of an adenovirus-SV40 recombinant producing SV40 T-antigen from adenovirus late promoter. Cell 24:135–143

Stamatatos L, Leventis R, Zuckermann MJ, Silivius JR (1988) Interaction of cationic lipid vesicles with negatively charged phospholipid vesicles and biological membranes. Biochemistry 27:3917–3925

Stein CA, Cheng Y-C (1993) Antisense oligonucleotides as therapeutic agents – is the bullet really magical? Science 261:1004–1012

Stopeck AT, Hersh EM, Akporiaye ET, Harris DT, Grogan T, Unger E, Warneke J, Schluter SF, Stahl S (1997) Phase I study of direct gene transfer of an allogeneic histocompatibility antigen, HLA-B7, in patients with metastatic melanoma. J Clin Oncol 15:341–349

Stull RA, Szoka FCJ (1995) Antigene, ribozyme and aptamer nucleic acid drugs: progress and prospects. Pharm Res 12:465–483

Sullenger BA, Cech TR (1993) Tethering ribozymes to a retroviral packaging signal for destruction of viral RNA. Science 262:1566–1569

Symons RH (1992) Small catalytic RNAs. Annu Rev Biochem 61:641–671

Symons RH (1994) Ribozymes. Curr Opin Struct Biol 4:322–330

Tang DC, Shi Z, Curiel DT (1997) Vaccination onto bare skin. Nature 388:729–730

Taylor NR, Kaplan BE, Swiderski T, Li H, Rossi JJ (1992) Chimeric DNA-RNA hammerhead ribozymes have enhanced in vitro catalytic efficiency and increased stability in vivo. Nucleic Acids Res 20:4559–4565

Temin HM (1995) Genetics of retroviruses. Ann NY Acad Sci 758:161–165

Templeton NS, Lasic DD, Frederik PM, Strey HH, Roberts DD, Pavlakis GN (1997) Improved DNA:liposome complexes for increased systemic delivery and gene expression. Nat Biotechnol 15:647–652

Thompson JD, Ayers DF, Malmstrom TA, McKenzie TL, Ganousis L, Chowrira BM, Couture L, Stinchcomb DT (1995) Improved accumulation and activity of ribozymes expressed from a tRNA-based RNA polymerase III promoter. Nucleic Acids Res 23:2259–2268

Thummel CR, Tjian R, Grodzicker T (1981) Expression of SV40 T antigen under control of adenovirus promoter. Cell 23:825–836

Tone T, Kashani-Sabet M, Funato T, Shitara T, Yoshida E, Kashfian BI, Horng M, Fodstad O, Scanlon KJ (1993) Suppression of EJ cell tumorigenicity. In Vivo 7:471–476

Tuschl T, Eckstein F (1993) Hammerhead ribozymes: importance of stem-loop II for activity. Proc Natl Acad Sci USA 90:6991–6994

Ueda K, Clark DP, Chen C-J, Roninson IB, Gottesman MM, Pastan I (1987) The human multidrug resistance (mdr1) gene. cDNA cloning and transcription initiation. J Biol Chem 262:505–508

Uhlenbeck OC (1987) A small catalytic oligoribonucleotide. Nature 328:596–600

Vahlsing HL, Yankauckas MA, Sawdey M, Gromkowski SH, Manthorpe M (1994) Immunization with plasmid DNA using a pneumatic gun. J Immunol Methods 175:11–22

van Beusechem VM, Kukler A, Heidt PJ, Valerio D (1992) Long-term expression of human adenosine deaminase in rhesus monkeys transplanted with retrovirus-infected bone marrow cells. Proc Natl Acad Sci USA 89:7640–7644

van Doren K, Hanahan D, Gluzman Y (1984) Infection of eukaryotic cells by helper-independent recombinant adenoviruses: early region I is not obligatory for integration of viral DNA. J Virol 50:606–614

van Tol H, Buzayan JM, Feldstein PA, Eckstein F, Bruening G (1990) Two autolytic processing reactions of a satellite RNA proceed with inversion of configuration. Nucleic Acids Res 18:19771–1975

Versnel MA, Hagemeijer A, Bouts MJ, van der Kwast TH, Hoogesteden HC (1988) Expression of c-sis (PDGF-B chain) and PDGF-A chain genes in ten human malignant mesothelioma cell lines derived from primary and metastatic tumors. Oncogene 2:601–605

Versnel MA, Claesson-Welsh L, Hammacher A, Bouts MJ, van der Kwast TH, Eriksson A, Willemsen R, Weima SM, Hoogesteden HC, Hagemeijer A, Heldin C-H (1991) Human malignant mesothelioma cell lines express PDGF-beta receptors whereas cultured normal mesothelioma cells express predominantly PDGF-alpha receptors. Oncogene 6:2005–2011

Vile RG, Hart IG (1993) In vitro and in vivo targeting of gene expression to melanoma cells. Cancer Res 53:962–967

Waddill W III, Wright W Jr, Unger E, Stopeck A, Akporiaye E, Harris D, Grogan T, Schluter S, Hersh E, Stahl S (1997) AJR Am J Roentgenol 169:63–67

Wagner E, Zatloukal K, Cotten M, Kirlappos H, Mechtler K, Curiel DT, Birnstiel ML (1992) Coupling of adenovirus to transferrin-polylysin/DNA complexes greatly enhances receptor-mediated gene delivery and expression of transfected genes. Proc Natl Acad Sci USA 89:6099–6103

Walz C, Schlehofer JR (1992) Modification of some biological properties of HeLa cells containing adeno-associated virus DNA integrated into chromosome 17. J Virol 66:2990–3002

Wei C-M, Gibson M, Spears PG, Scolnik EM (1981) Construction and isolation of a transmissible retrovirus containing the src gene of Harvey murine sarcoma virus and the thymidine kinase gene of herpes simplex virus type 1. J Virol 39:935–944

Weitzman MD, Wilson JM, Eck SL (1995) Adenovirus vectors in cancer gene therapy. In: Sobol RE, Scanlon KJ (eds) The internet book of gene therapy. Appleton and Lange, Stamford, pp 17–25

Werness BA, Levine AJ, Howley PM (1990) Association of human papilloma virus type 16 and 18 E6 protein with p53. Science 248:76–79

Whyte D, Buchkovic KJ, Horowitz JM, Friend SH, Raybuck M, Weinberg RA, Harlow E (1988) Association between an oncogene and an anti-oncogene: the adenovirus E1 A proteins bind to the retinoblastoma gene product. Nature 334:124–129

Williams RS, Johnston SA, Riedy M, DeVit JM, McEllington SG, Sanford JC (1991) Introduction of foreign genes into tissues of living mice by DNA-coated microprojectiles. Proc Natl Acad Sci USA 88:2726–2730

Woolf TM (1995) To cleave or not to cleave: ribozymes and antisense. Antisense Res Dev 5:227–232

Wright L, Wilson SB, Milliken S, Biggs J, Kearney P (1993) Ribozyme-mediated cleavage of the bcr/abl transcript expressed in chronic myeloid leukemia. Exp Hematol 21:1714–1718

Wu GY, Zhan P, Sze LL, Rosenberg AR, Wu CH (1994) Incorporation of adenovirus into a ligand-based DNA carrier system results in retention of original receptor specificity and enhances targeted gene expression. J Biol Chem 269:11542–11546

Yang NS, Burkholder J, Roberts B, Martinell B, McCabe D (1990) In vivo and in vitro gene transfer to mammalian somatic cells by particle bombardment. Proc Natl Acad Sci USA 87:9568–9572

Yang NS, Sun WH, McCabe D (1996) Developing particle-mediated gene-transfer technology for research into gene therapy of cancer. Mol Med Today 2:476–481

Yang Y, Ertle HC, Wilson JM (1994a) MHC class I-restricted cytotoxic T lymphocytes to viral antigens destroy hepatocytes in mice infected with E1-deleted recombinant adenoviruses. Immunity 1:433–442

Yang Y, Nunes FA, Berencsi K, Furth EE, Gohezd E, Wilson JM (1994b) Cellular immunity to viral antigens limits E1-deleted adenoviruses for gene expression. Proc Natl Acad Sci USA 91:4407–4411

Yu M, Ojwang J, Yamada O, Hampel A, Rappaport J, Looney D, Wong-Staal F (1993) A hairpin ribozyme inhibits expression of diverse strains of human immunodeficiency virus type 1. Proc Natl Acad Sci USA 89:6099–6103

Zakharchuk AN, Doronin KK, Karpov VA, Krougliak VA, Naroditsky BS (1995) The fowl adenovirus type 1 (CELO) virus-associated RNA-encoding gene: a new ribozyme-expression vector. Gene 161:189–193

Zamecnik PC, Stephenson ML (1978) Inhibition of Rous sarcoma virus replication and cell transformation by a specific oligodeoxynucleotide. Proc Natl Acad Sci USA 75:280–284

Zhao JJ, Pick L (1993) Generating loss-of-function genotypes of the fushi tarazo gene with targeted ribozyme in Drosophila. Nature 365:448–451

7 Cardiovascular Gene Therapy

R. Engler

7.1 Introduction

We are poised on the threshold of a new era; the ability to alter the genetic make up and/or expression of genes in humans. The prospect is a single injection of a gene therapeutic curing familial hypercholesterolemia, correcting severe combined immunodeficiency, removing the death sentence from those with a hereditary colon cancer gene, or growing new blood vessels to relieve chronic myocardial ischemia. The last decade has seen much progress toward reaching these goals, but we are not there yet. We review here the status of cardiovascular gene therapy in the published literature as of February 1998; mindful of the rapidly moving field we are surveying.

Gene therapy for the treatment of cardiovascular disease can be envisioned in several general ways. First, in congenital disease where there is failure to express a critical protein, or a mutant allele is expressed, the correct gene can be transfected. This strategy might require transfection of every cell within an organ system, or one specific cell type targeting (e.g., myocyte) when a structural protein is involved. When a metabolic or hormonal abnormality is present, only a subset of cells with less efficient targeting might be sufficient. Stable or long-term gene expression would be desirable. Second, in acquired diseases tissue remodeling could be achieved by transient expression of genes for paracrine or autocrine growth factors. The transfected gene might code for intracellular signaling molecules or transacting factors to induce cell division, cell death (apoptosis), cell differentiation, or a phenotype switch. Transient expression would be sufficient or even desirable. Examples include restenosis after angioplasty, scar formation after myocardial infarction, or arteriovenous malformations. Finally gene therapy might be used to alter a physiologic control system to correct an acquired defect. Optimal duration of expression and degree of targeting would depend on the circumstances. Examples include idiopathic hypertension, renal vascular hypertension, or altered cardiac β-adrenergic signal transduction in congestive heart failure. The desired targeting and duration of expression are a function of the disease and the vector used.

We review here the status of experimental gene therapy for cardiovascular disease from three points of view: the target cell, the disease target, and gene transfer technology.

7.2 Targeting Specific Cells or Organs

7.2.1 Myocardial Cells

Targeted cardiac expression with prolonged expression can be useful for correction of congenital or acquired disorders of myocytes, in which a gene transfer efficiency of 100% will be desirable. In treating conditions such as hypertrophic cardiomyopathy or reduced β-adrenergic responsiveness in congestive heart failure, the disorder exists in every cardiomyocyte, so highly efficient gene transfer will be required. In these nondividing cells incorporation of the transgene into the genome is not currently feasible and prolonged epichromosomal expression is required. In applications in which local expression of an apocrine hormone production can have a therapeutic effect, efficiency much lower that 100% may be effective. For cardiac apocrine applications, the cardiomyocyte need not be the target cell; equal efficacy might be achieved by targeting cardiac fibroblasts, cardiac endothelial cells, or cardiac vascular smooth muscle cells. Examples of apocrine effector gene therapy include growth factors for angiogenesis, growth hormone for congestive heart failure (CHF), and kallikrein or nitric oxide synthase for hypertension. In some applications systemic gene transfer without organ targeting could have highly undesirable effects.

Cardiac-specific promoters such as the α-myosin heavy chain can target cardiomyocytes. Transgenic mice using this targeted expression strategy have yielded important information about structural and functional proteins in the heart. However this strategy has not yet been applied to therapeutic cardiac targeting in disease models. Cell type-specific promoters have potential for generalized administration and organ-specific targeting, but expression level and degree of specificity need to be defined before application in humans. Cardiomyocytes can also be targeted by direct intramyocardial injection of adenovirus vectors or plasmid DNA, but expression is limited to the area of injection and significant local inflammatory reaction may be a significant clinical problem. Gene transfer efficiency with plasmid DNA is very low. With adenovirus, expression declines rapidly over several weeks in most tissues. Pericardial administration of adenovirus vectors results in epicardial gene transfer, and the expression is limited to the pericardium and epicardium. This might be useful for applications such as angiogenesis, however intrapericardial administration of angiogenic protein did not result in relief of ischemia in the dog model, suggesting that pericardial gene transfer for angiogenesis may not be effective. Both cardiac intramuscular and intrapericardial gene transfer require surgical intervention, or perhaps cardiac catheterization with specialized devices for injection.

Intracoronary administration of adenovirus in pigs was found to result in myocyte and endothelial transfection with high efficiency and relative targeting (Giordano et al. 1996). Surprisingly, over 98% of the adenovirus was taken up on the first pass through the heart, and in these pigs a serum factor partially neutralized the transfection efficiency of the virus that escaped the heart, yielding a very high degree of cardiac specificity. High first pass uptake may be due in part to high expression of αVβ3, αVβ5, and CAR (coxsackie-adenovirus receptor) or to the microcirculatory morphology in the heart, which is designed by nature for maximal blood/tissue exchange (Bergelson et al. 1997; Mathias et al. 1994). Adenoviral attachment occurs to cells expressing the coxsackie-adenovirus receptor, which is highly expressed in heart (Bergelson et al. 1997). Internalization is dependent on αVβ5 and, to a lesser extent, on αVβ3 integrins interacting with the penton base (Wickham et al. 1993, 1994). Once viral load is reduced by nearly 2 log units by passage through the heart, dilution of the virus in the total circulating blood volume and systemic distribution results in too few viral particles per cell for significant transfection. This strategy resulted in significant gene transfer of fibroblast growth factor (FGF)-5 to induce angiogenesis and completely relieved myocardial ischemia in the chronic pig ameroid model without any inflammatory reaction (Giordano et al. 1996).

7.2.2 Endothelial Cells and Vascular Smooth Muscle

Targeting of vascular endothelial cells has largely been accomplished by localized administration to restrict systemic distribution using a variety of devices and techniques. Hydrogel-coated balloons with plasmid DNA have been used experimentally and in patients for peripheral angiogenesis (Takeshita et al. 1996). Efficiency of transfection was low. Isolated segments of vessels can be transfected with cationic liposome/DNA complexes; adenovirus, naked DNA, and other vectors, but temporary interruption of the circulation to achieve dwell times of 10–20 min are required. Application of this strategy in critical vessels (e.g., coronary, carotid) will require temporary bypass or perfusion catheter placement through the isolated segment; both would seem to have limited practical application. Incubation of isolated vascular segments with adenovirus resulted in gene transfer to the endothelium and adventitia in normal vessels. This approach might be useful for vascular grafts. In the intact circulation the internal elastic lamina is a relative anatomic barrier to gene transfer, however limited transfer to smooth muscle and adventitia have been reported especially in diseased or injured vessels. Other strategies include high pressure injection, perfusion catheters, iontophoresis, perivascular injection, and intrapericardial delivery.

These limitations to gene transfer to vascular smooth muscle may be overcome using smooth muscle specific promoters. SM22-α is a tissue-specific vascular and visceral smooth muscle protein of unknown function. A 441 base pair (bp) *cis* element gives vascular smooth muscle-specific expression in culture when cloned upstream from lac-Z in an adenovirus vector. When injected intravenously in mice at 10^{10} pfu expression was limited to vascular smooth muscle in balloon-injured vessels, but when instilled into the bladder visceral smooth muscle was transfected. This promoter holds promise as a vascular smooth muscle targeting mechanism (Kim et al. 1997).

7.2.3 Cardiac Fibroblast

Transfection of cardiac fibroblasts by direct injection into myocardial granulation tissue of adenovirus vectors resulted in gene transfer (Murry et al. 1996). When the MYO-D gene in an adenovirus-5 vector driven by the Rous sarcoma virus (RSV) promoter was injected into the granulation tissue 7 days after myocardial necrosis, myofibroblasts displayed a phenotype switch to embryonic skeletal muscle. Fibroblast-specific promoters could also be used to increase specificity of gene transfer. This approach might be useful for gene transfer to granulation tissue to alter remodeling after acute myocardial infarction.

Table 1. Gene therapy for cardiovascular disease

Myocardial ischemia: chronic
Myocardial ischemia: acute
Limb ischemia
Restenosis
Inhibit vascular smooth muscle proliferation
Accelerate endothelial growth
Inhibit platelets, growth factors
Stents: inhibit stent restenosis
Hypercholesterolemia: LDL receptors
Thrombosis: enhance endothelial cell antithrombotic properties
Atherosclerosis: genetic disorders (e.g., homocysteineuria); enhanced expression of anti-atherogenic proteins
Transplant rejection
Transplant vascular disease

7.3 Cardiovascular Diseases

A list of conditions that might be amenable to gene therapy is given in Table 1. A brief review of the preclinical status of development follows.

7.3.1 Chronic Myocardial Ischemia

Angiogenesis for chronic myocardial ischemia has been studied in a number of laboratories. The most convenient model uses placement of an ameroid constrictor on a coronary artery which leads to gradual occlusion over 7–14 days and development of a collateral-dependent bed with minimal infarction. Initially collateral vessels are adequate for resting function, but fail to provide sufficient flow during stress or maximal vasodilatation. This model has been used extensively to test for angiogenic effects. Initially, protein therapy was tested. Protein therapy by intracoronary infusion with the growth factor basic FGF (bFGF) appears to require at least two infusions in the dog ameroid model; however, when studied after several months treated animals were no different than controls. The lack of long-term effect is likely a property of the dog model, in which the natural angiogenic response continues for perhaps 6 months until collateral vessels are sufficient during stress (e.g., the control animals "catch-up"). The dog model demonstrated that the natural angiogenic response can be accelerated by protein growth factor infusion, but this therapy has not been rigorously tested in a model in which the natural angiogenic process is insufficient for long-term relief of ischemia. In contrast, in pigs subjected to ameroid coronary occlusion, collateral vessel development is not sufficient; stable, stress-induced ischemia persists for at least 6 months (Roth et al. 1987, 1990). Gene therapy has been applied in several laboratories using the pig ameroid model. Direct epicardial injection into cardiac muscle of adenovirus transfecting the VEGF-121 gene relieved myocardial ischemia, and an Investigational New Drug Application (IND) for a Phase 1 safety trial has been initiated for application during coronary artery bypass surgery (Mack et al. 1998). Potential limitations include the necessity for general anesthesia, surgical delivery, local inflammatory reaction and the number of injection sites required for a beneficial response. However proof of principle, that the human heart can augment its angiogenic response, has recently been shown using direct acidic FGF protein injections during bypass surgery in humans (Schumacher et al. 1998). These findings lend support for development of either a gene- or protein-based angiogenic therapy. The method of delivery and production of the growth factor are largely a question of safety and convenience. Other strategies under development include endocardial injection with special catheter/needle devices and endocardial injection in conjunction with laser revascularization. Intracoronary injection using adenovirus-5/FGF-5 was highly effective in the pig model, and a commercial IND for a safety and efficacy trial using adenovirus-5 with the FGF-4 gene is underway. Several different genes from both the FGF and VEGF family have been found to be angiogenic in this model system.

Potential limitations or toxicity for angiogenic gene therapy include acceleration of atherosclerosis and potential to accelerate restenosis in patients who have had recent angioplasty. These potential complications would depend on circulation of the growth factor. This complication was found with protein infusion using VEGF-165 in a dog model (Lazarous et al. 1996). Circulation of proteoglycan-bound growth factors following cardiac-specific gene transfer is unknown. In peripheral vascular disease, when plasmid VEGF-165 was injected into skeletal muscle of patients with claudication and rest pain, the protein was found to circulate and result in significant bilateral leg edema (Isner et al. 1996). In theory gene transfer with an FGF family member might not have the edema effects of VEGF gene family members. VEGF is an endothelial-specific growth factor but also increases vascular permeability. FGF-4, FGF-5, VEGF-165, and many other family members are highly bound to the proteoglycan and, in response to injury in vivo, physiologic effects are localized by this property. Systemic effects and the tendency to circulate in patients will obviously depend on the amount of protein administered or produced from gene therapy. However, VEGF-121 is not proteglycan bound and would be more likely to circulate.

In summary, the normal biologic angiogenesis in response to ischemia is inadequate in patients who continue to have limb ischemia (see below) and/or myocardial ischemia. Augmentation of this natural response by gene therapy (or protein therapy) can result in further angiogenesis and relief of ischemia. It appears that a combination of the overproduction of a growth factor and an injury such as the ischemic environment is necessary for angiogenesis. However the hope that angiogenesis will be limited to an injury environment following gene or protein therapy is largely based on the absence of distant angiogenesis in animal models and has not been tested in humans. The observation that infused VEGF165 protein can accelerate stenosis following vascular injury in dogs indicates the need to look for accelerated atherosclerosis in clinical trials.

7.3.2 Acute Myocardial Ischemia

Several growth factors have been shown to reduce infarct size in organ ischemia and reperfusion models: bFGF is currently in clinical trial for reduction of morbidity and mortality following stroke; insulin-like growth factor (IGF)-1 and bFGF have been shown to reduce myocardial infarct size. Whether administration of these proteins or gene transfer to the heart by gene therapy can reduce infarct size and improve morbidity and mortality, have not been adequately tested in an animal model. Considerable preclinical work remains to be done before this type of therapy could be applied. Administration of growth factors in this setting is probably not effective through angiogenesis, but may prevent apoptosis. Two factors found to contribute to myocyte death in acute myocardial infarction are the inflammatory response and cardiomyocyte apoptosis; both might be altered by gene therapy. For example, cardiac-specific gene therapy to produce a local anti-inflammatory factor might decrease inflammatory damage. Gene transfer of a factor such as BCL-2 might limit cardiomyocyte apoptosis; acute expression might reduce infarct size and chronic expression might reduce the progression of CHF. This consideration is largely theoretical at the present time (Gottlieb et al. 1994).

Following myocardial infarction, granulation tissue is formed that leads to collagen scar. Both early and late remodeling of this tissue results in dilation, infarct expansion, and progression to CHF. Other factors may also contribute to progressive CHF such as continued cardiomyocyte apoptosis resulting in cell loss. Targeted gene therapy to the granulation tissue has the potential of altering scar formation. One exciting possibility is altering the phenotype of the myofibroblasts that lay down scar tissue following myocardial infarction. A feasibility study in a rat myocardial necrosis model used high dose adenoviral gene transfer with MYO-D and found myofibroblast phenotype switch to embryonic skeletal muscle (Murry et al. 1996). This experiment serves as a proof of principle that a phenotype switch with myofibroblasts is possible.

7.4 Peripheral Vascular Disease

7.4.1 Limb Ischemia

Angiogenic therapy with both protein and gene therapy have been found to improve limb ischemia in animal models (Yang et al. 1996; Tsurumi et al. 1996). As in the heart, the biology of collateral development by angiogenic proteins serves as a guide for gene therapy. Two reproducible reliable models (rat and rabbit femoral artery occlusion) have been developed. In the rat model continuous infusion of an angiogenic protein (bFGF) for up to 4 weeks results in continually improving blood flow over the entire 4 week period. When the protein infusion was stopped early and the animals followed, the improvement was about 50% of what was seen with 4 weeks of continuous therapy. Thus, sustained action of the growth factor for several weeks is necessary for a maximal response. These findings with protein therapy suggest that gene transfer to stimulate continuous growth factor production in the ischemic limb will have a significant advantage. Gene therapy with the variety of angiogenic growth factors has been found to relieve limb ischemia in the rabbit model (Tsurumi et al. 1996; Takeshita et al. 1996; Asahara et al. 1995).

Initial gene transfer technology utilized plasmid DNA and a hydrogel-coated balloon to achieve gene transfer to the endothelial cell layer of the arterial wall. This therapy was initially applied in patients with ischemia at rest and threatened limb loss. While initial results were encouraging, it appeared that gene transfer was limited and intramuscular injection of plasmid DNA with VEGF-165 was substituted. The results in a small number of patients in an uncontrolled clinical trial suggest a remarkable improvement in functional capacity and brachial ankle index (Isner et al. 1996). In this trial, nearly all of the patients had significant edema requiring diuretic therapy, perhaps as a consequence of the vascular permeability effect of VEGF. These results await confirmation in a properly performed placebo-controlled and blinded clinical trial. They do provide, however, a solid foundation for the relative safety and potential improvement possible in patients with severe limb ischemia.

7.4.2 Restenosis

Restenosis occurring in the first 3 months following percutaneous transluminal coronary angioplasty is a significant clinical problem occurring in 30%–40% of patients. A number of animal models have been developed in an attempt to mimic the clinical condition. However these models have not been reliable in predicting the results of subsequent clinical trials. It may be that models that induce cellular proliferation do not mimic the pathophysiology of restenosis at all and that inhibiting proliferation could worsen plaque instability (Libby 1998). With that in mind, three different approaches have been used in restenosis models to inhibit smooth muscle proliferation. The first involves direct inhibition of smooth muscle proliferation through genes that cause the death of cells entering the cell cycle. For example, gene transfer with the herpes simplex thymidine kinase gene to areas of vascular injury, followed by administration of ganciclovir at the appropriate time, reduces the proliferate response (Guzman et al. 1994). Additional experimental gene transfer to inhibit vascular smooth muscle proliferation include nitric oxide synthase, GAX (growth arrest homeobox), retinoblastoma gene, p21 and antisense gene transfer for the cyclins and cyclin-dependent kinases. A second approach is to use an endothelial-specific growth factor to accelerate endothelial regrowth following vascular injury. Use of VEGF gene in an animal model resulted in significant improvement over control (Asahara et al. 1996). A third strategy would be gene transfer to inhibit platelets or growth factor production locally. Virtually all of the gene therapy strategies require prolonged exposure of the vascular wall to the gene therapy technique utilized. Long dwell times in temporarily isolated vascular segments are likely to have limited clinical utility in coronary or carotid arteries. One approach to this problem is antibody receptor targeting. The goal is to inject a gene therapy vector for systemic circulation and target the cell type using cell-specific receptors. VEGF receptors are expressed nearly exclusively on endothelial cells and could provide an antigen that specifically identifies endothelial cells. Adenovirus with antibody attached could target the receptor and facilitate cell type-specific gene transfer or shorten dwell time following local administration. Critical factors in developing a gene therapy treatment will be efficiency and timing of gene transfer and site-specific localization. Deployment of endovascular stents has shown significant success in preventing restenosis. Nonetheless, even stents have some incidence of smooth muscle proliferation leading to stenosis. Gene therapy product-coated stents might be useful to accelerate re-endothelialization of the stent and/or prevent smooth muscle growth.

7.4.3 Hypercholesterolemia

The most direct approach to treat familial hypercholesterolemia will be targeted gene transfer for the low density lipoprotein (LDL) receptor. Pre-clinical studies in the Watanabe hypercholesterolemic rabbit model have demonstrated dramatic reductions in LDL cholesterol and an increase in high density lipoprotein (HDL) (Li et al. 1995). However, using the E1-deleted adenovirus injected into the splenic vein, gene expression was transient and the cholesterol levels returned to baseline within 3 weeks of gene transfer. The problem of transient expression using the first generation adenovirus vector is being approached by further deletion of viral genes, or ex vivo gene transfer with retrovirus for stable incorporation into the genome followed by cell implantation. Use of autologous hepatocytes will require ex vivo cell culture. Considerable genetic manipulation will be required to achieve efficient hepatic gene transfer. Other monogenetic disorders as potential targets include apolipoprotein E deficiency and lipoprotein lipase deficiency.

A second approach would be to limit the atherosclerotic process. Some experimental success has been seen with gene transfer of the nitric oxide synthase (NOS) gene to vascular endothelium. Augmenting NOS production could have a variety of beneficial effects including platelet inhibition, neutrophil and macrophage inhibition, and vasodilatation. Furthermore, nitric oxide signaling has been found to be an important downstream messenger in VEGF-induced angiogenesis;, however, the potential angiogenic effect of gene transfer with endothelial nitric oxide synthase (eNOS) has not been tested.

7.4.4 Hypertension

The etiology and mechanism of essential hypertension in the vast majority of patients remain an enigma. Gene therapy designed to alter a specific etiologic gene is therefore only a future hope. Nonetheless, several animal model experiments using gene therapy led to reduced blood pressure. Adenovirus gene transfer of the human kallikrein gene by intravenous or intramuscular delivery significantly lowered blood pressure in the spontaneous hypertensive rat (SHR), two kidney one clip Goldblatt, and Dahl salt-sensitive hypertensive rats (Chao and Chao 1997). The blood pressure lowering effect was prevented by administration of the kallikrein-specific antagonist aprotinin, indicating that kallikrein mediated the effect. Gene transfer with cytomegalovirus (CMV)-eNOS plasmid DNA injected into the tail vein of SHR rats lowered blood pressure for up to 12 weeks (Lin et al. 1997). While these experiments provide only symptomatic lowering of blood pressure, they might be quite useful until mechanism-specific therapy is available.

7.4.5 Thrombosis

Circulating blood exists in a state of constant balance between thrombosis and thrombolysis. When this balance is shifted in pathologic disease states, arterial and/or venous thrombosis becomes a common pathologic mechanism in a number of congenital and acquired diseases. In atherosclerosis, plaque instability leading to acute myocardial infarction usually involves thrombosis. For example, abnormalities in coagulation proteins S and C as well as other less common genetic defects also lead directly to venous thromboembolic disease or may predispose to arterial thrombosis and complicate atherosclerosis. Gene transfer targeted to endothelial cells, where the fluid/thrombus interface is often formed, or targeted to liver cells responsible for producing coagulation proteins could potentially accomplish reduction in the tendency for intravascular thrombosis. Several experimental strategies for enhanced anti-thrombotic protein production by endothelial cells have been developed. For example, prostacyclin is an important paracrine factor inhibiting vascular thrombosis and platelet adhesion. In a porcine model of carotid angioplasty, gene transfer with adenovirus-COX-1 (cyclooxygenase 1) inhibited cyclic flow variations (a platelet-mediated phenomenon) and thrombosis (Zoldhelyi et al. 1996). Gene transfer required a 30 min intravascular dwell time, a problem that might be overcome with high pressure injections or other techniques. In a rabbit iliac angioplasty/stent model of vascular injury, gene transfer with plasmid VEGF-165 by hydrogel balloon accelerated re-endothelialization and reduced thrombosis (Van Belle et al. 1997). Other potential antithrombotic genes include NOS, plasminogen activators, and other antithrombotic proteins. Defective coagulation due to abnormal proteins such as C or S could be accomplished by insertion into the genome of hepatocytes or ectopic production by direct injection into skeletal muscle. In applications in which the circulating level of a protein is critical in maintaining a balance between thrombosis and thrombolysis, promoters responsive to administration of pharmacologic agents might be used to control expression.

7.4.6 Atherosclerosis

A number of genetic risk factors for atherosclerosis have been identified. Specific genetic disorders such as homocysteinuria and familial hypercholesterolemia could be corrected by gene therapy.

7.4.7 Transplant Rejection

Despite considerable progress, transplant rejection and immunosuppressive drug therapy to prevent rejection remain significant problems. One interesting possibility for gene therapy to prevent rejection has come from study of the phenomenon of immune privilege. In areas such as the eye, lack of immune response had been thought to occur by exclusion of immune competent cells from the microcirculation and/or the extravascular areas of these organs. Recent evidence suggests that expression of Fas ligand is a significant factor (Griffitz et al. 1995; Bellgrau et al. 1995). The Fas receptor on the cell surface is linked to death domain binding proteins that initiate apoptosis, or programmed cell death. For example, clonal deletion of self-reactive immune-competent T cells is signaled through Fas ligand binding to Fas on the T cells. In an attempt to utilize this strategy for transplantation, an intriguing study recently reported that cotransplant of myotubes expressing Fas ligand with pancreatic β-cells resulted in prolonged survival of the donor cells (Lau et al. 1996). This observation has potential application in organ transplant, where ex vivo perfusion to accomplish gene transfer to express Fas ligand by a variety of techniques (e.g. adenovirus, adeno-associated virus, cationic liposomes, etc.) could provide protection from an immune response. A number of technical problems and fundamental questions must be answered in the basic research laboratory before this potentially exciting technique can be applied.

7.5 Gene Transfer Technology

Targeting gene transfer or gene expression to an organ or cell type and duration of expression are two critical problems for cardiovascular gene therapy. Several of the gene transfer techniques utilized in cardiovascular gene therapy applications have been discussed in earlier sections, and we will review these techniques here with emphasis on targeting and duration of expression. First generation (E1 gene-deleted) adenovirus is particularly suited to situations in which transient expression is desirable. One important application is apocrine hormone-mediated effects of growth factors for angiogenesis. In this situation local production of the growth factor is desirable and results in new blood vessel growth. Once the blood vessels are formed, continuing flow maintains their structure and the initiating growth factor is no longer required; once the highway is built, the construction crew can leave. In some applications prolonged expression is desired. With the goal of prolonging expression, mechanisms of loss of expression with E1-deleted adenovirus have been investigated and two components have been identified. The first may involve primary viral load that results in death of the adenovirus-transfected cell. This has been shown to occur in a number of situations and is related to the multiplicity of infection (MOI). MOIs of 1000 viral particles per cell, or higher, result in toxicity and cell death. MOIs between 10 and 100 are usually not directly toxic to the cell. After transfection and production of transgene protein, immune mechanisms become important. Small amounts of viral proteins appear to be produced despite E1 deletion, and these will be displayed by the transfected cell depending on its ability to function as an antigen-presenting cell. Display of these viral protein results in immune response (Yang et al. 1994; Yang and Wilson 1995) as evidenced by antibody formation and T cell responses. Cells such as hepatocytes that display viral antigens after E1-deleted adenovirus transfection may be killed by CD8 lymphocytes (Yang et al. 1994). However the display and lymphocyte response is dependent on the cell type and its capability as an antigen-presenting cell. Cytotoxic lymphocyte reactions have not been seen after intracoronary transfection with first generation adenovirus in both pig and rabbit models, perhaps related to ability to present antigens. Furthermore, decreases in production of transgene protein without death of the transfected cell occur, and these are sensitive to anti-inflammatory treatment with cyclosporin A. This may be related to a second immune mechanism of diminishing protein production over time. Here the immunogenicity of the transgene is involved. For example in experiments using a second generation adenoviral vector two transgenes were studied in parallel gene transfer experiments in mice. One transgene was the murine (self) and the other was the human (foreign) erythropoietin gene. Erythropoetin levels and hemoglobin levels demonstrated prolonged gene expression of the murine gene but markedly shorter expression of the human gene. Furthermore, expression of foreign transgene is prolonged in immune incompetent mice or rats (Schumacher et al. 1998; Yang et al. 1995; Smith et al. 1996). In other experiments it was confirmed that prior immunization reduces gene transfer efficiency and immune suppression can enhance or prolong gene expression (Schulick et al. 1997). Several second generation adenovirus vectors have been created by deletion of additional early genes E2, E3, or E4 in order to further limit viral protein production and immune reaction. Several of these vectors have been shown to prolong transgene expression by limiting the immune response to viral protein (Yang et al. 1994). However, the multi-deleted vectors have limitations in production that must be overcome. One overarching principle in all the adenovirus gene transfer experiments is that route and/or site of administration are a critical determinant of immune reaction. For example, intramyocardial injection results in inflammation and fibrosis whereas intracoronary injection does not (Giordano et al. 1996; Barr et al. 1994).

Gene transfer for immunization, or production of a circulating hormone, (e.g., erythropoietin to increase red blood cell production), can be at any convenient site. However, targeted transfection or expression is important in some applications. One interesting approach to cell-specific transfer is to target cell surface receptors. Either agonists for cell surface receptors or antibodies directed against a cell specific surface marker can be linked to plasmid DNA or to recombinant adenovirus. For cardiovascular transfer to endothelial cells the VEGF receptor might be used. Targeted expression has been shown using FGF family receptors (Sosnowski et al. 1996).

Cationic liposomes for transfer of plasmid DNA have been used in isolated arterial segments and by direct injection into muscle. In arterial segments, prolonged dwell times are required, as with adenovirus. Following intramuscular injection, using a variety of enhancing techniques, gene transfer is still significantly less than the efficiency seen with adenovirus.

Targeted gene transfer to arterial segments has generally required up to 20 min of dwell time, which limits practical application during noninvasive therapy. In the case of adenovirus, the prolonged exposure time may be in part due to extremely slow diffusion times. There are a number of potential approaches to solving the exposure time problem in arterial gene transfer. Poloxamer 407 has been shown to reduce required dwell time for gene transfer by over 50% and to increase transfection efficiency by increasing vascular permeability. Other techniques for increasing vascular permeability, such as administration of VEGF protein, might facilitate gene transfer.

Adeno-associated virus (AAV) has potential as a very useful tool when prolonged expression is required. Production problems with AAV have been at least partially solved and availability of AAV with marker genes has increased preclinical research activity using this vector. Similar to adenovirus, intracoronary delivery of AAV also results in cardiac gene transfer without inflammation, but the efficiency was much less than what was achieved with adenovirus (Kaplitt et al. 1996). This strategy has not yet been applied in a cardiovascular disease model. Use of cell-specific promoters such as myosin heavy chain or SM22 was discussed above.

The hemagglutinating virus of Japan (HVJ), a Sendai virus, has been used to enhance gene transfer with cationic lipid/DNA complexes. Liposomes are prepared and mixed with HVJ inactivated by UV irradiation. Intramuscular injection results in gene transfer but with the same local inflammatory reactions as adenovirus. Intracoronary injection can achieve low level gene transfection, mostly in perivascular and vasa vasorum areas, without inflammation. It would seem to have no advantage over adenovirus (Aoki et al. 1997).

When prolonged or indefinite expression is desirable, retrovirus gene transfer may be useful. These vectors incorporate into the genome of dividing cells and will thus be present in daughter cells. Stable transfection into endothelial cells in culture has been demonstrated (Zwiebel et al. 1989). This strategy could be used with ex vivo transfection and seeding of transgenic cells. The extensive literature on retroviral gene transfer techniques is beyond the scope of this review.

7.5.1 Human Cardiovascular Gene Therapy Trials

Cardiovascular gene therapy trials in humans have been limited to date. A pioneering peripheral vascular disease trial using plasmid DNA with

VEGF-165 has achieved some success, but lacked placebo controls (Isner et al. 1996). Collateral Therapeutics and Berlex Biosciences have initiated a trial of angiogenesis for coronary artery disease in patients with stable clinical angina pectoris. At Cornell University in New York, a clinical trial of direct injection of adenovirus transfer of the VEGF-121 for angiogenesis in patients undergoing coronary artery bypass graft surgery has been initiated. In Finland both cationic liposomes and adenovirus with the VEGF-165 transgene is being injected intracoronary. No results are yet available for the cardiac gene transfer trials.

In conclusion, cardiovascular gene therapy is an area of intense investigative activity at the current time. There is tremendous promise for potential to treat a number of cardiovascular diseases with this technique. After 10 years of preclinical research, the first applications will be tested in the clinic in a variety of diseases over the next few years. While problems may arise that were not anticipated in the preclinical work, we have a strong theoretical and experimental basis, and the time has come to proceed with clinical trials.

References

Aoki M, Morishita R, Muraishi A, Moriguchi A, Sugimoto T, Maeda K, Dzau VJ, Kaneda Y, Higaki J, Ogihara T (1997) Efficient in vivo gene transfer into the heart in the rat myocardial infarction model using the HVJ (hemagglutinating virus of Japan)-liposome method. J Mol Cell Cardiol 29:949–959

Asahara T, Bauters C, Zheng LP, Takeshita S, Bunting S, Ferrara N, Symes JF, Isner JM (1995) Synergistic effect of vascular endothelial growth factor and basic fibroblast growth factor on angiogenesis in vivo. Circulation 92:II365–371

Asahara T, Chen D, Tsurumi Y, Kearney M, Rossow S, Passeri J, Symes JF, Isner JM (1996) Accelerated restitution of endothelial integrity and endothelium-dependent function after phVEGF165 gene transfer. Circulation 94:3291–3302

Barr E, Carroll J, Kalynych AM, Tripathy SK, Kozarsky K, Wilson JM, Leiden JM (1994) Efficient catheter-mediated gene transfer into the heart using replication-defective adenovirus. Gene Ther 1:51–58

Bellgrau D, Gold D, Selawry H, Moore J, Franzusoff A, Duke RC (1995) A role for CD95 ligand in preventing graft rejection (see comments). Nature 377:630–632

Bergelson JM, Cunningham JA, Droguett G, Kurt-Jones EA, Krithivas A, Hong JS, Horwitz MS, Crowell RL, Finberg RW (1997) Isolation of a common receptor for coxsackie B viruses and adenoviruses 2 and 5. Science 275:1320–1323

Chao J, Chao L (1997) Experimental kallikrein gene therapy in hypertension, cardiovascular and renal diseases. Pharmacol Res 35:517–522

Giordano FJ, Ping P, McKirnan MD, Nozaki S, DeMaria AN, Dillmann WH, Mathieu-Costello O, Hammond HK (1996) Intracoronary gene transfer of fibroblast growth factor-5 increases blood flow and contractile function in an ischemic region of the heart (see comments). Nat Med 2:534–539

Gottlieb RA, Burleson KO, Kloner RA, Babior BM, Engler RL (1994) Reperfusion injury induces apoptosis in rabbit cardiomyocytes. J Clin Invest 94:1621–1628

Griffith TS, Brunner T, Fletcher SM, Green DR, Ferguson TA (1995) Fas ligand-induced apoptosis as a mechanism of immune privilege. Science 270:1189–1192

Guzman RJ, Hirschowitz EA, Brody SL, Crystal RG, Epstein SE, Finkel T (1994) In vivo suppression of injury-induced vascular smooth muscle cell accumulation using adenovirus-mediated transfer of the herpes simplex virus thymidine kinase gene. Proc Natl Acad Sci USA 91:10732–10736

Isner JM, Pieczek A, Schainfeld R, Blair R, Haley L, Asahara T, Rosenfield K, Razvi S, Walsh K, Symes JF (1996) Clinical evidence of angiogenesis after arterial gene transfer of phVEGF165 in patient with ischaemic limb. Lancet 348:370–374

Kaplitt MG, Xiao X, Samulski RJ, Li J, Ojamaa K, Klein IL, Makimura H, Kaplitt MJ, Strumpf RK, Diethrich EB (1996) Long-term gene transfer in porcine myocardium after coronary infusion of an adeno-associated virus vector. Ann Thorac Surg 62:1669–1676

Kim S, Lin H, Barr E, Chu L, Leiden JM, Parmacek MS (1997) Transcriptional targeting of replication-defective adenovirus transgene expression to smooth muscle cells in vivo. J Clin Invest 100:1006–1014

Lau HT, Yu M, Fontana A, Stoeckert CJ (1996) Prevention of islet allograft rejection with engineered myoblasts expressing FasL in mice. Science 273:109–112

Lazarous DF, Shou M, Scheinowitz M, Hodge E, Thirumurti V, Kitsiou AN, Stiber JA, Lobo AD, Hunsberger S, Guetta E, Epstein SE, Unger EF (1996) Comparative effects of basic fibroblast growth factor and vascular endothelial growth factor on coronary collateral development and the arterial response to injury. Circulation 94:1074–1082

Li J, Fang B, Eisensmith RC, Li XH, Nasonkin I, Lin-Lee YC, Mims MP, Hughes A, Montgomery CD, Roberts JD (1995) In vivo gene therapy for hyperlipidemia: phenotypic correction in Watanabe rabbits by hepatic delivery of the rabbit LDL receptor gene. J Clin Invest 95:768–773

Libby P (1998) Gene therapy of restenosis. Circ Res 82:404–406 (abstract)

Lin KF, Chao L, Chao J (1997) Prolonged reduction of high blood pressure with human nitric oxide synthase gene delivery. Hypertension 30:307–313

Mack CA, Patel SR, Schwarz EA, Zanzonico P, Hahn RT, Ilercil A, Devereux RB, Goldsmith SJ, Christian TF, Sanborn TA, Kovesdi I, Hackett N, Isom OW, Crystal RG, Rosengart TK (1998) Biologic bypass with the use of adenovirus-mediated gene transfer of the complementary deoxyribonucleic acid for vascular endothelial growth factor 121 improves myocardial perfusion and function in the ischemic porcine heart. J Thorac Cardiovasc Surg 115:168–176; discussion 176–177

Mathias P, Wickham T, Moore M, Nemerow G (1994) Multiple adenovirus serotypes use alpha v integrins for infection. J Virol 68:6811–6814

Murry CE, Kay MA, Bartosek T, Hauschka SD, Schwartz SM (1996) Muscle differentiation during repair of myocardial necrosis in rats via gene transfer with MyoD. J Clin Invest 98:2209–2217

Roth DM, Maruoka Y, Rogers J, White FC, Longhurst JC, Bloor CM (1987) Development of coronary collateral circulation in left circumflex ameroid-occluded swine myocardium. Am J Physiol 253:H1279–H1288

Roth DM, White FC, Nichols ML, Dobbs SL, Longhurst JC, Bloor CM (1990) Effect of long-term exercise on regional myocardial function and coronary collateral development after gradual coronary artery occlusion in pigs. Circulation 82:1778–1789

Schulick AH, Vassalli G, Dunn PF, Dong G, Rade JJ, Zamarron C, Dichek DA (1997) Established immunity precludes adenovirus-mediated gene transfer in rat carotid arteries. Potential for immunosuppression and vector engineering to overcome barriers of immunity. J Clin Invest 99:209–219

Schumacher B, Pecher P, von Specht BU, Stegmann T (1998) Induction of neoangiogenesis in ischemic myocardium by human growth factors. Circulation 97:645–650

Smith TA, White BD, Gardner JM, Kaleko M, McClelland A (1996) Transient immunosuppression permits successful repetitive intravenous administration of an adenovirus vector. Gene Ther 3:496–502

Sosnowski BA, Gonzalez AM, Chandler LA, Buechler YJ, Pierce GF, Baird A (1996) Targeting DNA to cells with basic fibroblast growth factor (FGF2). J Biol Chem 271:33647–33653

Takeshita S, Tsurumi Y, Couffinahl T, Asahara T, Bauters C, Symes J, Ferrara N, andIsner JM (1996) Gene transfer of naked DNA encoding for three isoforms of vascular endothelial growth factor stimulates collateral development in vivo. Lab Invest 75:487–501

Takeshita S, Weir L, Chen D, Zheng LP, Riessen R, Bauters C, Symes JF, Ferrara N, Isner JM (1996) Therapeutic angiogenesis following arterial gene transfer of vascular endothelial growth factor in a rabbit model of hindlimb ischemia. Biochem Biophys Res Commun 227:628–635

Tsurumi Y, Takeshita S, Chen D, Kearney M, Rossow ST, Passeri J, Horowitz JR, Symes JF, Isner JM (1996) Direct intramuscular gene transfer of naked DNA encoding vascular endothelial growth factor augments collateral development and tissue perfusion (see comments). Circulation 94:3281–3290

Van Belle E, Tio FO, Chen D, Maillard L, Kearney M, Isner JM (1997) Passivation of metallic stents after arterial gene transfer of phVEGF165 inhibits thrombus formation and intimal thickening. J Am Coll Cardiol 29:1371–1379

Wickham TJ, Filardo EJ, Cheresh DA, Nemerow GR (1994) Integrin alpha v beta 5 selectively promotes adenovirus mediated cell membrane permeabilization. J Cell Biol 127:257–264

Wickham TJ, Mathias P, Cheresh DA, Nemerow GR (1993) Integrins alpha v beta 3 and alpha v beta 5 promote adenovirus internalization but not virus attachment. Cell 73:309–319

Yang HT, Deschenes MR, Ogilvie RW, Terjung RL (1996) Basic fibroblast growth factor increases collateral blood flow in rats with femoral arterial ligation. Circ Res 79:62–69

Yang Y, Ertl HC, Wilson JM (1994) MHC class I-restricted cytotoxic T lymphocytes to viral antigens destroy hepatocytes in mice infected with E1-deleted recombinant adenoviruses. Immunity 1:433–442

Yang Y, Nunes FA, Berencsi K, Furth EE, Gonczol E, Wilson JM (1994) Cellular immunity to viral antigens limits E1-deleted adenoviruses for gene therapy. Proc Natl Acad Sci USA 91:4407–4411

Yang Y, Nunes FA, Berencsi K, Gonczol E, Engelhardt JF, Wilson JM (1994) Inactivation of E2a in recombinant adenoviruses improves the prospect for gene therapy in cystic fibrosis. Nat Genet 7:362–369

Yang, Y, Trinchieri G, Wilson JM (1995) Recombinant IL-12 prevents formation of blocking IgA antibodies to recombinant adenovirus and allows repeated gene therapy to mouse lung (see comments). Nat Med 1:890–893

Yang Y, Wilson JM (1995) Clearance of adenovirus-infected hepatocytes by MHC class I-restricted CD4+ CTLs in vivo. J Immunol 155:2564–2570

Zoldhelyi P, McNatt J, Xu XM, Loose-Mitchell D, Meidell RS, Clubb FJ, Buja LM, Willerson JT, Wu KK (1996) Prevention of arterial thrombosis by adenovirus-mediated transfer of cyclooxygenase gene. Circulation 93:10–17

Zwiebel JA, Freeman SM, Kantoff PW, Cornetta K, Ryan US, Anderson WF (1989) High-level recombinant gene expression in rabbit endothelial cells transduced by retroviral vectors. Science 243:220–222

8 Protection of Hematopoietic Progenitor Cells from Chemotherapy Toxicity by Transfer of Drug Resistance Genes

N. Takebe, S.-C. Zhao, D. Banerjee, and J.R. Bertino

8.1 Approaches to Modify Chemotherapy Induced Myelosupression

A major obstacle in eradicating unresectable or disseminated tumor cells with chemotherapy is thought to be due to inherent drug resistance, which in some instances may be overcome by the use of higher doses (Hryniuk and Bush 1984). Currently available chemotherapy, however, is often associated with various toxicities, in particular, bone marrow toxicity which is limiting for many of these agents. Use of recombinant hematopoietic growth factors or cytokines has enabled a modest increase in the dose chemotherapy that is tolerated. Most commonly used cytokines are granulocyte colony-stimulating factor (G-CSF) and granulocyte macrophage colony-stimulating factor (GM-CSF). Clinical trials using interleukin-3 (IL-3) and PIXY 321 (fusion protein of GM-CSF and IL-3) did not show a significant advantage over control groups (Ellis et al. 1996; Hofstra et al. 1997). Thrombopoietin, alone or in combination with other cytokines, is currently being evaluated in clinical trials. Another means of overcoming myelosuppression is the use of autologous bone marrow or peripheral stem cell transplantation after high doses of chemotherapy. This approach is currently being used in Hodgkin's disease, breast and ovarian carcinomas (Bierman et al. 1993; Rosti et al. 1992; Reece et al. 1992; Chopra et al. 1993; Barnett et al. 1993).

Transfer of drug resistance genes into hematopoietic cells is a promising approach as a method to reduce induced myelosuppression. Interestingly, transfer of drug resistance genes into hematopoietic precursors protects not only bone marrow cells but seems to reduce other related toxicities such as gastrointestinal (GI) toxicity if granulocyte counts remain high during the post-chemotherapy period (May et al. 1995; Li et al. 1994).

8.2 Methods for Gene Transfer into Hematopoietic Progenitor Cells

Viral vectors have been the method of choice for introducing foreign genes into hematopoietic progenitor cells (Fig. 1). In particular, retroviral vectors have been preferred as these vectors show stable integration into the host genome and produce long-term gene expression. Nonviral vectors such as liposomes have been evaluated from a safety point of view (Walker et al. 1996) as well as for introducing foreign genes into localized solid tumors, but, thus far, low transduction efficiency limits their use for hematopoietic cells. Other viral vectors such as adenovirus, adeno-associated virus (AAV), and human immunodeficiency virus (HIV) are also being tested as vectors for hematopoietic gene transfer (Kotin et al. 1990; Zhou et al. 1992; Chatterjee et al. 1992). Adenovirus was reported as being incapable of infecting bone marrow cells, but recently investigators have reported successful infection of hematopoietic progenitor cells under high multiplicity of infection (MOI) conditions (Huang et al. 1995). Adenovirus does not integrate stably into the host genome, thus duration of gene expression is relatively short. Lentiviruses (e.g., HIV) can infect quiescent cells and thus hold promise for gene transfer into nondividing cells; they are currently under investigation (Naldini et al. 1996).

8.2.1 Retrovirus-Mediated Gene Transfer

Retroviral gene transfer strategies involve use of packaging lines to generate replication incompetent virus particles. Packaging cell lines contain unique sequences which have a defective packaging signal, but contain gag, pol, and env signals to produce viral proteins. The packaging cell line is transfected with the retroviral vector plasmid, which contains the packaging signal and the gene of interest. As a consequence, transfected viral RNA is encapsulated by viral proteins generated by packaging cell lines and virus containing the gene of interest is secreted. This virus can be used to infect target cells and, after reverse transcription, is integrated into the target cell genome (Gilboa et al. 1986). To prevent the generation of replication-competent helper virus, newer packaging cell lines have been designed that require at least three recombination events to occur before producing replication-competent retrovirus (Markowitz et al. 1988a,b). There have been no reports of helper virus production using these packaging cell lines.

Hematopoietic stem cell gene transfer efficiency depends upon two important factors, the titer of the viral vector and the cell cycle status of the target cells. The viral titer is determined by the packaging cell line

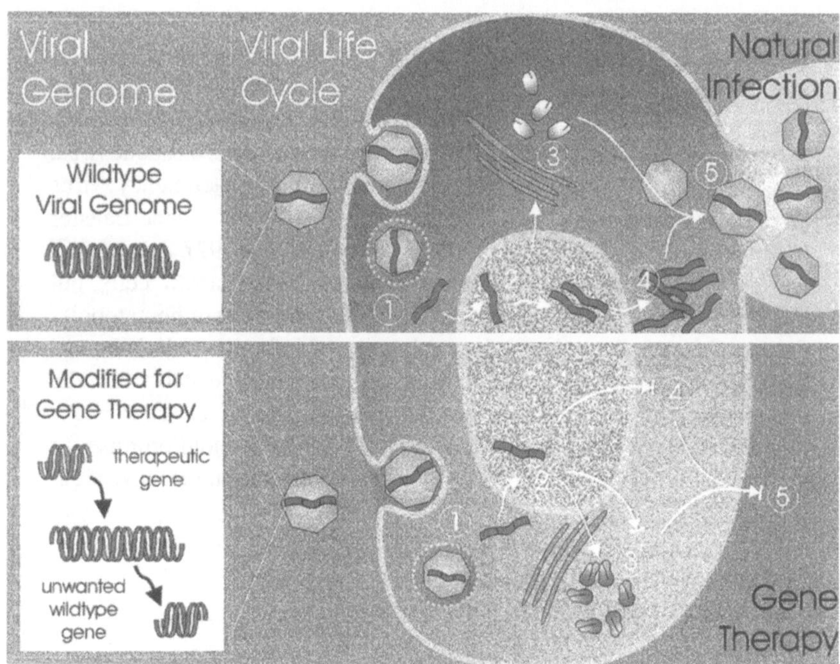

Fig. 1. Overview of an ideal gene therapy virus vector. Both naturally occurring and genetically altered virus enter the cell by receptor-mediated endocytosis (①), and their genetic material becomes incorporated into the nucleus (②). In naturally occurring infection (*top*), the virus initiates transcription and expression of its genetic information in the host cell, manufacturing the viral proteins (③) and replication of the viral genome (④), which finally leads to assembly of infectious virus particles and disruption of the host cell membrane. Subsequently the viral infection spreads to other cells (⑤). For gene therapy (*bottom*), harmful viral genes for protein expression and replication have been genetically altered to be no longer detrimental to the cell; while a gene coding for a therapeutic protein has been inserted instead. Thus, the therapeutic viral vector initiates the expression of the desired therapeutic gene product (*red*), while no other viral proteins are produced (③), no virus genome is replicated (④), and thus, no further spread of infection is possible (⑤)

quality and the sequence of the retroviral construct. Efforts have been made to improve packaging cell lines as well as retroviral constructs. The length of the gag region in the packaging cell line improves the packaging efficiency and increases the titer, while the sequences of the inserted gene determine the stability of the vector genome and packaging efficiency (Bender et al. 1987; Armentano et al. 1987). The majority of hematopoietic stem cells in bone marrow are thought to be in the G_0 state and are thus difficult targets for retrovirus infection (Ogawa 1989). To bring quiescent bone marrow stem cells into cycle, a combination of growth factors have been administered with some improvement in transduction efficiency (Bodine et al. 1989). To achieve the maximum expression level of the gene of interest in hematopoietic cells, the viral LTR (long terminal repeat) contains promoter and enhancer elements which may improve expression levels (Bowtell et al. 1987). The myeloproliferative sarcoma virus (MPSV) enhancer is known to increase transduced gene expression in myeloid cells (Bowtell et al. 1988) and is under investigation.

Mouse stem cell transduction studies using the adenosine deaminase (ADA) gene, glucocerebrosidase gene, or β globin gene showed successful gene transfer and expression of the gene of interest in CFU-S colonies and long-term expression in reconstituted mice (Wilson et al. 1990; Nolta et al. 1990; Dzierzak et al. 1988). A retroviral marking study in humans using the neomycin phosphotransferase gene into marrow or peripheral blood CD34+ cells showed a low level of efficiency and only short-term persistence of the marker gene (Brenner et al. 1993; Dunbar et al. 1995). In addition to the strategies to improve transduction efficiency discussed so far, modification of transduction protocols such as the use of stromal cells or, more recently, fibronectin are being explored (Dao et al. 1995; Hanenberg et al. 1995). Virus infections are carried out using viral supernatants or by coculture. Although the latter results in higher transduction efficiency (Bodine et al. 1991), there is a substantial loss of stem cells due to adherence to the viral producer cells. Recently, use of vesicular stomatitis virus (VSV) G protein instead of the retroviral envelope has been reported (Burns et al. 1993). This pseudotyped virus G protein has a higher affinity for phosphatidylserine, which exists in the target cell membrane, and increases transduction efficiency and viral receptor-mediated transfer (Burns et al. 1993).

8.2.2 Preparation of the Target Hematopoietic Stem Cells

While isolation of the true stem cell remains elusive, purification methods using commercially available columns or immunomagnetic beads enrich for CD34+ progenitor cells (Moore et al. 1996). In order to stimulate stem cells into cycle, cytokines such as kit ligand or stem cell factor (SCF), IL-1, IL-3, and IL-6 are used (Ogawa 1989; Bodine et al. 1989; Moore and Hoskins 1994; Moore 1995; Sha et al. 1995). Actual transduction methods reported by different investigators range from simple transduction to those requiring stromal support or long-term marrow culture (Moore et al. 1992). Cord blood stem cells have also been used for gene transfer experiments. Ex vivo expansion of transduced progenitor cells may increase the number of transduced cells (Moore and Hoskins 1994). Recently, successful hematopoietic engraftment after high dose chemotherapy using ex vivo expansion of autologous bone marrow cells by stromal-based bioreactors has been reported. This new technology offers a promising approach for future gene transfer experiments, although further research on the long-term repopulation ability of expanded CD34+ of peripheral blood or cord blood will be required (Stiff et al. 1997).

8.3 Drug Resistance Genes

Genes expressing proteins which confer drug resistance are candidates for gene transfer to protect the bone marrow from the toxicity of the drug (Banerjee et al. 1994). To be applicable in the clinic, these genes must be well characterized with regard to the effects of their expression in target cells, especially those which are down-regulated during maturation of progenitor cells. Continuous expression of such genes in differentiated cells may affect normal physiological function of these cells.

8.3.1 O^6-Alkylguanine-DNA Alkyltransferase

0^6-alkylguanine-DNA-alkyltransferase (alkyltransferase) protein is an enzyme with a molecular weight of 18,000–26,000, responsible for repair of 0^6-alkylguanine adducts which are formed in cells exposed to nitrosoureas (Pegg et al. 1983). 0^6-alkyl (methyl and ethyl) guanine seems to be the predominant premutagenic lesion which causes mispairing, leading to point mutations (Pegg 1984). 0^6-alkylguanine is removed from DNA by alkyltransferase to a cysteine residue within its active site, leaving an intact guanine base in DNA (Pegg 1984; Pegg et al. 1983). As a result, this protein is thought to decrease the cytotoxic and mutagenic effects of DNA damaging agents (Tong et al. 1982; Ludlum et al. 1986) and to protect cells from antitumor drugs and carcinogen-induced malignant transformations (Dolan et al. 1990; Brennand and Margisos 1986). A unique aspect of this protein is that it reacts stoichiometrically, which means that each protein molecule is capable of transferring one alkyl group from 0^6-alkylguanine to the protein with a half-life of between 15 and 47 min (Pegg et al. 1983). The protein becomes permanently inactivated during this process. Thus, the repair capability for 0^6-alkylguanine is influenced by the level of active alkyltransferase molecules. Several tissues lack the 0^6-alkylguanine adduct repair system due to low levels of the DNA repair protein (Pegg et al. 1983; Gerson et al. 1987). In hematopoietic precursors, alkyltransferase is expressed at low levels (Gerson et al. 1985, 1987, 1996). Thus myelosuppression is the dose-limiting toxicity of nitrosoureas. The combination of 0^6-benzylguanine (0^6-BG), which inactivates this enzyme in tumor cells as well as normal tissue, plus N,N'-bis(2-chloroethyl)-N-nitrosourea (BCNU), now in clinical trials, while sensitizing tumor cells containing this enzyme, also leads to myelosuppression (Gerson et al. 1996).

Transfection of a plasmid containing the bacterial alkyltransferase into mammalian cells conferred resistance to chlorozotocin and nitrogen mustard (Brennand and Margisos 1986). In another set of experiments, NIH-3T3 cells did not show increased resistance to chloroethylnitrosoureas (BCNU) in spite of increased alkyltransferase activity after gene transfer, although in CCL-1 cells, both increased bacterial alkyltransferase activity and the resistance to BCNU were demonstrated (Dumenco et al. 1989). Transfer of bacterial alkyltransferase by electroporation into murine hematopoietic stem cells which are alkyltransferase deficient showed expression of alkyltransferase. These cells became resistant to methylating agents but not to nitrogen mustards (Jelinek et al. 1988). Transfection of human methylguanine-DNA methyltransferase (MGMT) cDNA (Pegg 1990; Tano et al. 1990) into Chinese hamster cells, which lack alkyltransferase activity, made the alkyltransferase expressing clones resistant to some of the alkylating agents and caused a reduction in mutation frequency induced by other alkylating agents (Kaina et al. 1991). From the observations that tumor cells expressing high levels of alkyltransferase are resistant to killing by methylating or chloroethylating drugs, work to synthesize compounds that block the alkyltransferase activity to enhance their action was initiated (Crone and Pegg 1993). The proline at position 140 in mammalian alkyltransferase was replaced by alanine, which resulted in resistance to 0^6-benzylguanine, an inactivator of 0^6-alkylguanine-DNA alkyltransferases and a potent inhibitor of the mammalian alkyltransferases (Dolan et al. 1990). In addition to the proline residue at position 140, mutations at position 138 and 156 reduce the ability of the enzyme to react with O^6-benzylguanine, and, expectedly, a combination of these mutations rather than a single mutation produced greater resistance to 0^6-benzylguanine (Crone et al. 1994). Almost simultaneously, two different groups demonstrated that retroviral transduction and expression of the human alkyltransferase cDNA conferred nitrosourea resistance to hematopoietic cells (Allay et al. 1995; Moritz et al. 1995).

A myeloproliferative sarcoma virus (MPSV) based retrovirus (vM5MGMT) was used to express the human alkyltransferase cDNA in K562 cells and murine bone marrow progenitors (Allay et al. 1995). Transduced K562 cells showed high levels of alkyltransferase expression and increased (threefold) resistance to BCNU in comparison to uninfected K562 cells. G418 selected clones showed a 21-fold increase in resistance to BCNU as compared to the untransduced or cells. Murine BM cells were transduced with retrovirus containing the *MGMT* construct; however, they showed increased resistance to BCNU by only 1.4-fold. In bone marrow transplantation experiments in mice, *MGMT* transduced cells were shown to express the alkyltransferase gene, as demonstrated by RT-PCR amplification and western blot analysis. Another study compared the resistance to BCNU in an in vivo mouse model using both mock transduced and virally transduced cells. Although deaths occurred in both arms after 6 weeks of drug treatment, animals receiving *MGMT* transduced cells showed less leukopenia, less thrombocytopenia and anemia (Moritz et al. 1995). In vivo mice experiments demonstrated that the severity of anemia and thrombocytopenia but not leukopenia were reduced in mice treated with 0^6-BG and BCNU by overexpression of the bacterial alkyltransferase gene. A clear survival advantage was conferred by alkyl transferase gene transfer in the mice treated with 12.5 mg/kg BCNU but there was no significant difference in the survival of the groups at other doses (7.5 mg/kg, 10 mg/kg,15 mg/kg).

More recently, a human alkyltransferase cDNA in a retrovirus construct was introduced into murine primary bone marrow cells (Jelinek et al. 1996). These transduced hematopoietic progenitors (CFU-GM) conferred three times more resistance to MNU (N-methyl-N-nitrosourea) than the progenitors present in control cultures. In the long-term culture system, the output of progenitor cells was much less affected by repeated exposure to MNU, although the control culture demonstrated a steady decline of progenitor cells. To overcome the toxic effect of bone marrow progenitors from 0^6-BG, mutated human alkyltransferase, with the proline at position 140 changed to an alanine and an additional mutation of glycine at position 156 to alanine, was introduced into Chinese hamster lung fibroblast (RJKO) cells. There was no protective effect against 0^6-chloroethylguanine adducts, but in the presence of 0^6-BG, cells expressing mutated alkyltransferase were resistant to mitozolamide and chlorozotocin (Hickson et al. 1996). Future studies using retroviral-mediated gene transfer of the mutated human alkyltransferase into human bone marrow progenitor cells will be of interest.

8.3.2 MDR-1

The multi-drug resistance gene (*mdr*-1) encodes a 1280-amino acid protein called P-glycoprotein (P170) expressed in the plasma membrane of some normal and some tumor cells (Pastan and Gottesman 1987; Gottesman and Pastan 1988; Endicott and Ling 1989). This multidrug transporter is an ATP-dependent protein that effluxes a variety of drugs from cells to maintain the intracellular levels of these drugs at nontoxic levels (Kartner and Ling 1989; Deuchars and Ling 1989). The cDNAs for the *mdr*-1 gene from mouse (Gros et al. 1986), as well as human (Roninson et al. 1986; Gros et al. 1986) cells have been cloned and characterized. P-glycoprotein expression was initially thought to be absent in peripheral and bone marrow cells. However, P-glycoprotein is highly expressed in pluripotent stem cells but decreases in expression as the hematopoietic progenitors undergo maturation (Chaudhary and

Roninson 1991; Drach et al. 1992; Chaudhary et al. 1992). A transgenic mouse model expressed the *mdr*-1 transgene at all stages of hematopoietic maturation in the modified cells at higher levels than unmodified cells (Galski et al. 1989). Bone marrow cells from these mice were resistant to the cytotoxic effect to MDR drugs. These studies have led investigators to utilize this gene for marrow gene replacement therapy in humans (Pastan and Gottesman 1991). The human *mdr*-1 gene was introduced into mouse bone marrow cells and the transduced cells showed resistance to both colchicine and vinblastine (McLachlin et al. 1990). Furthermore, in murine bone marrow transplantation experiments using *mdr*-1 transduced bone marrow cells, an enrichment for transduced cells was observed when transplanted mice were treated with taxol (Sorrentino et al. 1992). Other investigators have transduced this *mdr*-1 cDNA into mouse erythroleukemia cells (Delaflor-Weiss et al. 1992) and human CD34+ cells (Ward et al. 1994) by retroviral-mediated transfer. Serial bone marrow transplantation of irradiated mice displayed an increased level of MDR-1 expression and resistance to taxol treatment supporting the notion that sufficient numbers of early precursor cells were transduced by *mdr*-1 containing virus (Hanania and Deisseroth 1994). An aberrant splicing of *mdr*-1 transcripts in producer clones resulting in reduced expression of vector-derived P-glycoprotein has been reported (Hesdorffer et al. 1994). In current clinical trials designed to increase marrow resistance to cytotoxic drugs, the *mdr*-1 gene has been used in treating breast and ovarian carcinoma (O'Shaughessy et al. 1994; Deisseroth et al. 1994; Hesdorffer et al. 1994, 1997). Successful *mdr*-1 gene transduction with stromally transduced cells but not with suspension transduced cells in breast and ovarian carcinoma patients has been observed (Kavanagh et al. 1996). Results of another Phase I clinical trial indicated successful bone marrow cell engraftment but low transduction efficiency in two patients with positive PCR signals for *mdr* (Hesdorffer et al. 1997). Further work needs to be carried out in order to determine if patients receiving *mdr*-1 gene-transduced stem cells can actually tolerate higher curative doses of taxol. Table 1 includes a list of approved and ongoing clinical trials using *mdr*-1 cDNA.

Table 1. Currently approved clinical protocols for transfer of drug resistance genes into human hematopoietic cells

Gene	Vector	Target	Trans-duction	Post-transplant therapy	Disease (reference)
hMDR-1	HaMSV	PBMC/BM	S	P, V	Breast (O'Shaughnessy et al. 1994)
hMDR-1	MoMLV	BM CD34+	S, auto stroma	P	Breast, ovarian (Deisseroth et al. 1994)
hMDR-1	HaMSV	BM CD34+	S	P	Breast, ovarian glioma (Hessdorfer et al. 1994, 1997)

MDR, multi-drug resistance; HaMSV, Harvey murine sarcoma virus; MoMLV, Moloney murine leukemia virus; S, supernatant; P, paciltaxel; V, vinblastine.

8.3.3 Cytidine Deaminase

The cDNA for cytidine deaminase (CD), a 146-amino acid protein, has been cloned and expressed (Laliberte and Momparler 1994). This enzyme catalyzes the conversion of several cytidine analogue cytotoxic drugs to nontoxic metabolites, including cytosine arabinoside (Ara-C), 5-azacytidine, and 2',2'-difluorodeoxycytidine (gemcitabine). Based on the hypothesis that overexpression of this enzyme might confer cellular resistance to cytidine analogues, two different groups transferred human cytidine deaminase cDNA into murine bone marrow cells and CEM cells (Momparler et al. 1996; Neff and Blau 1996). Murine hematopoietic cell clonogenic assays showed 90% survival at an Ara-C concentration of 10^{-6} M vs no survival in controls (Momparler et al. 1996). In a CEM cell line, transduced cells showed 2.1-fold increased resistance to Ara-C (Neff and Blau 1996).

8.3.4 Aldehyde Dehydrogenase Class 1

Aldehyde dehydrogenase (ALDH) catalyzes the conversion of aliphatic and aromatic aldehydes to the corresponding acids. In humans, five ALDH isozymes have been purified and characterized. ALDH 1 and

ALDH 3 were reported to be overexpressed in cyclophosphamide resistant cell lines (Hilton 1984; Yoshida et al. 1993; von Eitzen et al. 1994; Sreerama and Sladek 1994; Rekha et al. 1994; Hsu et al. 1992). Among all the isozymes, cytosolic ALDH 1 has been suggested to be involved in the cellular cyclophosphamide resistance (Hilton 1984; Russo and Hilton 1988). ALDH 1 is the only isozyme known to be expressed in immature bone marrow progenitor cells and its level decreases with hematopoietic differentiation which correlates with decreased cyclophosphamide resistance (Kastan et al. 1990). Overexpression of this enzyme in hematopoietic progenitors may confer cyclophosphamide resistance.

The cDNA for human ALDH 1 has been cloned and characterized (Hsu et al. 1985; Zheng et al. 1993) and is 1.5 kb in length. The active form of cyclophosphamide, 4-hydroxycyclo-phosphamide (4HC), is formed by hydroxylation of cyclophosphamide by cytochrome p450. After cellular uptake of 4-HC, its open chain form, aldophosphamide, generates the alkylating agents acrolein and phosphoramide mustard. ALDH oxidizes aldophosphamide into nontoxic carboxyphosphamide.

ALDH 1 cDNA in a retroviral vector has been introduced into a human monoblastic leukemia cell line and murine hematopoietic cell lines which showed 1.5-to 3-fold resistance to maphosphamide, an active analogue of cyclophosphamide (Magni et al. 1996). Transduction of human peripheral blood hematopoietic progenitors showed resistance to maphosphamide as well. Moreb et al. (1996) have observed resistance to 4-hydroperoxy-cyclophosphamide in K562 cells transduced by ALDH 1 cDNA.

8.3.5 Dihydrofolate Reductase

Folate analogs such as methotrexate (MTX) target dihydrofolate reductase (DHFR), which catalyzes conversion of folate and dihydrofolate to tetrahydrofolate, a cofactor in the biosynthesis of purines and thymidylate (Blakley 1984). One of the mechanisms of resistance to MTX observed in cultured cells are mutations in the *DHFR* gene leading to an enzyme with reduced affinity for antifolates. The first mutant form of DHFR described, the Arg-22 mutation, was demonstrated to act as a dominant selectable marker in transfected cultured mammalian cells (Simonsen and Levinson 1983). Several DHFR mutants have been identified subsequently, with increased catalytic efficiency and decreased MTX binding, and have been utilized both for in vitro and in vivo gene transfer studies (Isola and Gordon 1986; McIvor and Simonsen 1991; Hussaon et al. 1992; Banerjee et al. 1992, 1994; Williams et al. 1987; Corey et al. 1990; Li et al. 1994; Schweitzer et al. 1989; Ercikan-Abali et al. 1996; Nakahara et al. 1997). In 1987, Williams et al. showed that murine bone marrow cells transfected with a mutant *DHFR* and transplanted into lethally irradiated recipients showed a survival advantage over a nontransduced group when challenged with MTX (Williams et al. 1987). Newer mutant human DHFRs with increased catalytic efficiency and decreased MTX binding include Phe-3l, try22, Ser-3l, (Alt et al. 1978; Pizzorno et al. 1989) Phe-22 (Dicker et al. 1990), Trp-31 (McIvor and Simonsen 1990). Recently mutants with dual substitutions at both Leu-22 and Phe-3l resulting in Phe-22/Ser3l, Phe-22/Gly-3l, Tyr-22/Ser-3l, and Tyr-22/Gly-3l have also been generated (Ercikan-Abali et al. 1996), and these show very high levels of resistance to MTX and trimetrexate (TMTX).

In our laboratory, the Ser-3l mutant has been extensively studied (Banerjee et al. 1992; Li et al. 1994; Zhao et al. 1994). Irradiated mice transplanted with bone marrow cells infected with this vector showed a survival advantage over control mice when treated with high doses of MTX. In mice bearing a transplanted mammary carcinoma, transduction of marrow with this vector allowed mice to tolerate curative doses of MTX, and five of ten animals had complete and lasting tumor regression (Zhao et al. 1996). The CD34+ subset of human peripheral blood and cord blood stem cells have also been infected with the Ser-3l human *DHFR* retroviral construct. The infected CD34+ population was threefold more resistant to MTX than the noninfected population (Flasshove et al. 1995). Recently we evaluated different mutants of DHFR with respect to levels of resistance to MTX. NIH3T3 cells infected by retroviral vectors containing either wild-type (wt) or mutant *DHFR* cDNA showed significant increases of IC_{50} for the double mutant F22S31 by 28-and 850-fold over the wt transfected and nontransfected cells (Nakahara et al. 1997). The Y22 mutated *DHFR* gene also confers 100-fold higher resistance to TMTX in murine hematopoietic progenitor cells (Spencer et al. 1996). Bone marrow transplantation using the Y22 mutant showed hematopoietic protection against TMTX-induced neutropenia in mice. Mice transplanted with marrow cells from transgenic mice expressing mutated DHFRs (either the Arg-22 or Trp-31) demonstrated improved survival against MTX toxicity over animals transplanted with normal marrow (May et al. 1995).

8.3.6 Glutathione Associated Enzymes

Among the glutathione associated enzymes, glutathione S-transferase (GST) plays a role in drug inactivation resulting in anticancer drug resistance (Becket and Hayes 1993). There are four major isoforms of cytosolic GST enzymes, α,μ, π, and θ. Among these, the α and μ forms are associated with cellular resistance to nitrogen mustard and nitrosoureas (McGown and Fox 1986; Tew 1994). Retrovirus-mediated gene transfer of the rat *Yc* gene (GSTα family) into 3T3 mouse fibroblast cells resulted in a five-to tenfold increase in resistance to alkylating agents (Greenbaum et al. 1994). CFU-GM assays using GST-π transduced human CD34+ cells in the presence of 4-hydroxy-cyclophosphamide (4-HC) showed resistance to 4-HC as compared to nontransduced cells (Kuroda et al. 1996).

8.3.7 Bcl-2

Initially discovered because of its involvement in the t(14;18) translocations found in follicular small cell and diffuse large cell lymphomas (Tsujimoto et al. 1985), the *bcl*-2 (B-cell lymphoma/leukemia-2) protooncogene inhibits apoptotic cell death. The *bcl*-2 gene is located at 18q21 in juxtaposition with the immunoglobulin heavy chain locus at 14q32, resulting in its transcriptional deregulation (Tsujimoto et al. 1985; Reed et al. 1989). Apoptosis is blocked by *bcl*-2 in in vitro models involving a variety of external stimuli such as hyperthermia, radiation, growth factor withdrawal, glucocorticoids, and multiple classes of chemotherapeutic agents (Miyashita and Reed 1992, 1993; Naumovski and Cleary 1994). Hockenbery et al. (1993) demonstrated that cells overexpressing *bcl*-2 still generate peroxidases, but do not damage their cellular constituents including lipid membranes. Further evidence comes from the finding that *bcl*-2 deficient mice develop two potentially redox related pathologies, polycystic kidney disease and hypopigmentation (Veeis et al. 1993).

Recently, it has been shown that *bcl*-2 is part of a higher molecular weight complex (Oltvai et al. 1993). One of the first *bcl*-2 related genes to be discovered is *bax*. It heterodimerizes with *bcl*-2 and this complex is thought to regulate cell survival or death following exposure to an apoptotic stimulus depending on the ratio of the two genes (Hoffman and Leibermann 1994). Another *bcl*-2 associated gene is *bcl*-x, which codes for two distinct mRNAs generated by alternative splicing (Boise et al. 1993). The resulting two different proteins have opposite effects when transfected into cells. Similar in size to *bcl*-2, *bcl*-xl was shown to inhibit cell death as effectively as *bcl*-2 in an IL-3-dependent cell line. In comparison, *bcl*-xs, encoding for a smaller protein, was found to inhibit the ability of *bcl*-2 to enhance the survival of the IL-3 deprived cells. Thus, *bcl*-xs acts like bax. In contrast to *bax*, *bcl*-xs or *bcl*-xl was not found to dimerize with *bcl*-2. Transfection of *bcl*-2 cDNA into mammalian cells results in resistance to a variety of chemotherapeutic agents (Miyashita and Reed 1992, 1993). Kondo et al. (1994) reported that retrovirus-mediated gene transfer of the *bcl*-2 gene into bone marrow cells protected mice from chemotherapy induced myelosuppression, without compromising the antitumor activity of the administered chemotherapeutic agents.

8.3.8 Manganese-Superoxide Dismutase

The superoxide dismutases are a class of metalloproteins that catalyze the dismutation of superoxide radicals to oxygen and hydrogen peroxide. Their presence in all aerobic life and absence from anaerobic life suggests that these enzymes play a critical role in protecting cells against oxidative stress (Wong et al. 1989). It has been demonstrated that high levels of manganese-superoxide dismutase (Mn-SOD) are induced in several tumor cell lines as well as in normal cells in vitro and in vivo by treatment with IL-1 or tumor necrosis factor (TNF). Overexpression of Mn-SOD in tumor cells which were transfected by Mn-SOD cDNA promoted cell survival after exposure to IL-1, TNF, ionizing radiation or chemotherapeutic agents such as mitomycin C and doxorubicin (Wong et al. 1989; Hirose et al. 1993).

8.4 Drug Resistance Genes as Selectable Markers

One of the major limitations of gene transfer into hematopoietic progenitors is the difficulty in achieving large numbers of transduced cells. Exceptions are limited to successful gene transfer of genes such as human adenosine deaminase in humans, a deficiency of which only requires a small number of target blood cells to be corrected (Bordignon et al. 1995). Introduction of genes into hematopoietic progenitor cells to correct hemoglobinopathies and metabolic storage diseases require a large number of target cells to be successfully transfected. Drug resistance genes may be used as selectable markers in gene therapy to increase the population of transduced genes (Sha et al. 1995).

8.4.1 In Vivo Selection

The *mdr*-1, *DHFR* and human methylguanine methyltransferase (*MGMT*) genes are good candidates for selectable markers. Sorrentino et al. (1992) demonstrated increased proviral DNA copy numbers in circulating leukocytes obtained from mice which received a bone marrow transplant using *mdr*-1 transduced cells and taxol treatment. Others have confirmed these observations at the protein and functional level by demonstrating an increased percentage of leukocytes positive for P-glycoprotein (Podda et al. 1992) and by the rhodamine efflux assay (Davis et al. 1996). After transduction of a mutated human *DHFR* gene into mouse bone marrow cells, these cells were returned to lethally irradiated mice, followed by low dose to high dose MTX treatment. Primary, secondary and tertiary recipients with transduced marrow showed an increase in MTX resistant CFU-CM colonies (Li et al. 1994; Zhao et al. 1994). In mice reconstituted with bone marrow cells containing a mutated *MGMT* gene, increased resistance to BCNU was observed compared to mice transplanted with wild-type *MGMT* transduced marrow cells (Davis et al. 1996).

8.4.2 Selection Ex Vivo

Another approach to increase the transduced cell population is to select for transduced cells ex vivo before reinfusing the cells to the host. This approach spares the host from the toxic effects of chemotherapy drugs and provides a means for generating a larger proportion of transduced cells. G418 was used to attempt selection in cells transduced with the Neo^R gene, but this treatment decreased the repopulating ability of murine hematopoietic tissue grafts (Dick et al. 1985; Kaleko et al. 1990). G418 may not be an ideal selection agent because Neo^R gene expression may be toxic to engrafted hematopoietic cells (Apperley et al. 1991). *DHFR* transduced peripheral blood progenitor cells made resistant to MTX can be selectively expanded ex vivo, with MTX added in culture, and more than twofold expansion of transduced cells over untransduced cells has been observed in experimental systems (Flasshove et al. 1995). *mdr*-1 transduced cells have been isolated using an anti-P-glycoprotein antibody resulting in an increased proportion of MDR-1 expressing cells in transplanted mice (Richardson and Bank 1995). Colchicine has also been used as a pretransplant selection drug and increased MDR-1 expression in transplanted mice was observed (Aran et al. 1994; Licht et al. 1995).

8.5 Summary

Several genes have been studied and are capable of rendering hematopoietic progenitor cells resistant to chemotherapy. Successful gene transfer of hematopoietic cell precursors has the promise of allowing safe administration and administration of even higher doses of chemotherapy than are now used in the clinic, possibly resulting in an increased in cures of some tumors. Drug resistance genes may also be useful as dominant selectable markers to introduce other genes of interest into marrow precursors, especially those that cannot be selected, e.g., β globin. Viral vectors that are able to infect nondividing stem cells and are also integrated into the target cell genome will be the ideal vectors for stem cell gene transfer. Chimeric vectors incorporating useful characteristics of both adenoviral and retroviral vectors have recently been described and may become useful for transduction of marrow progenitor cells (Feng et al. 1997). Another promising vector system is the attenuated lentiviral vector system derived from HIV-1, which infects nondividing cells and can be used for long-term gene expression (Zufferey et al. 1997). With improvements in vector-designed transduction efficiencies, transfer of drug resistance genes into hematopoietic stem cells will become an important means of protecting bone marrow from toxicity associated with high dose chemotherapy.

References

Allay JA, Dumenco LL, Koc ON, Liu L, Gerson SL (1995) Retroviral transduction and expression of the human alkyltransferase cDNA provides nitrosourea resistance to hematopoietic cells. Blood 85:3342

Alt FW, Kellems RE, BertinoJR, Schimke RT (1978) Selective amplification of dihydrofolate reductase genes in methotrexate resistant variants of cultured murine cells. J Biol Chem 253:1357

Apperley JF, Luskey BD, Williams DA (1991) Retroviral gene transfer of human adenosine deaminase in murine hematopoietic cells. Effect of selectable marker sequences on long-term expression. Blood 78:310

Aran JM, Gottesman MM, Pastan I (1994) Drug-selected coexpression of human glucocerebroside and P-glycoprotein using a bicistronic vector. Proc Natl Acad Sci USA 91:3176

Armentano D, Yu SF, Kantoff PW, von Ruden T, Anderson WF, Gilboa E (1987) Effect on internal viral sequences on the utility of retroviral vectors. J Virol 61:1647

Banerjee D, Schweitzer BI, Volkenandt M et al (1992) Transfection with a cDNA encoding a Ser3 l or Ser34 mutant human dihydrofolate reductase into Chinese Hamster Ovary and mouse marrow progenitor cells confers methotrexate resistance. Gene 139:269

Banerjee D, Zhao SC, Tong Y, Steinherz J, Gritsman K, Bertino JR (1994a) Transfection of a nonactive site mutant murine DHFR cDNA (the Tryptophan 15 mutant) into chinese hamster ovary and mouse marrow progenitor cells imparts MTX resistance in vitro. Cancer Gene Ther l:181

Banerjee D, Zhao SC, Li M-X, Schweitzer BI, Mineishi S, Bertino JR (1994b) Genetherapy utilizing drug resistance genes: a review. Stem Cells 12:378

Barnett MJ, Coppin CM, Murray N, Nevill TJ, Reece DE, Klingemann HG, Shepherd JD, Nantel SH, Sutherland HJ, Phillips GL (1993) High-dose chemotherapy and autologous bone marrow transplantation for patients with poor prognosis nonseminomatous germ cell tumors. Br J Cancer 68:594

Becket GJ, Hayes JD (1993) Glutathione S-transferases; biomedical applications. Adv Clin Chem 30:281

Bender NU, Palmer TD, Gehnas RE, Miller AD (1987) Evidence that the packaging signal of Moloney murine leukemia virus extends into the gag region. J Virol 61:1639

Bertino JR (1996) Active site-directed double mutants of dihydrofolate reductase. Cancer Res 56:4142

Bierman PJ, Bagin RG, Jagannath S, Vose JM, Spitzer G, Kessinger A, Dicke KA, Armitage JO (1993) High dose chemotherapy followed by autologous hematopoietic rescue in Hodgkin's disease: long-term follow-up in 128 patients. Ann Oncol 4:767

Blakley RL (1984) Dihydrofolate Reductase. In: Blakley RL, Benkovic SJ (eds) Folates and pterins, vol 1. Wiley, New York, pp 191–253

Bodine DM, Karlsson S, Nienhuis AW (1989) Combination of interleukins 3 and 6 preserves stem cell function in culture and enhances retrovirus-mediated gene transfer into hematopoietic stem cells. Proc Natl Acad Sci USA 86:8897

Bodine DM, McDonagh KT, Seidel NE, Nienhuis AW (1991) Survival and retrovirus infection of murine hematopoietic stem cells in vitro: effects of 5-FU and method of infection. Exp Hematol 19:206

Boise LH, Gonziaz-Garcia M, Postema CE, Ding L, Lindster T, Turka LA, Mao X, Nunez G, Thompson CB (1993) Bcl-x, a bcl-2-related gene that functions as a dominant regulator of apoptotic cell death. Cell 74:597–608

Bordignon C, Notarangelo LD, Nobili N, Ferrari G, CasoratiG, Panina P, Mazzorari N, Maggioni D, Rossi C, Servida P, Ugazio AG, Mavillo F (1995) Gene therapy in peripheral blood lymphocytes and bone marrow for ADA immunodeficient patients. Science 270:470

Bowtell DDL, Johnson GR, Kelso A, Cory S (1987) Expression of genes transferred to hematopoietic stem cells by recombinant retroviruses. Mol Biol Med 4:229

Bowtell DDL, Cory S, Johnson GR, Gonda TJ (1988) Comparison of expression in hematopoietic cells by retroviral vectors carrying two genes. J Virol 62:2464

Brennand J, Margisos GP (1986) Reduction of the toxicity and mutagenicity of alkylating agents in mammalian cells harboring the Escherichia coli alkyltransferase gene. Proc Natl Acad Sci USA 83:6292

Brenner MK, Rill DR, Holaday MS et al (1993) Gene marking to determine whether autologous marrow infusion restores longterm hematopoiesis in cancer patients. Lancet 342:1134

Burns JC, Friedmann T, Driever W et al (1993) Vesicular stomatitis virus G glycoprotein pseudotyped retroviral vectors: concentration to very high titer and efficient gene transfer into mammlian and nonmammalian cells. Proc Natl Acad Sci USA 90:8033–8037

Chatterjee S, Johnson PR, Wong KK (1992) Dual-target inhibition of HIV-1 in vitro by means of adeno-associated virus antisense vector. Science 258:1485

Chaudhary PM, Roninson IB (1991) Expression and activity of P-glycoprotein, a multidrug efflux pump, in human hematopoietic stem cells. Cell 66:85–94

Chaudhary PM, Mechetner EB, Roninson IB (1992) Expression and activity of the multidrug resistance P-glycoprotein in human peripheral blood lymphocytes. Blood 80:2735

Chopra R, McMillan AK, Linch DC, Yuklea S, Taghipour G, Pearse R, Patterson KG, Goldstone AH (1993) The place of high-dose BEAM therapy and autologous bone marrow transplantation in poor-risk Hodgkin's disease. A single-center eight-year study of 155 patients. Blood 81:1137

Corey CA, DeSilva AD, Holland CA, Williams DA (1990) Serial transplataion of methotrexate resistant bone marrow; protection murine recipients from drug toxocity by progeny of transduced stem cells. Blood 76: 337

Crone T, Goodtzova K, Edara S et al (1994) Mutations in human 0^6-alkylguanine-DNA alkyltransferase imparting resistance to 0^6-benzylguanine. Cancer Res 54:6221–6227

Crone TM, Pegg AE (1993) A single amino acid change in human 0^6-alkylguanine-DNA alkyltransferase decreasing sensitivity to inactivation by 0^6-benzylguanine. Cancer Res 53:4750–4753

Dao MA, Hannum C, Nolta JA et al (1995) Flt3 ligand preserves viability of human hematopoietic stem cells during in vitro transduction (abstract). Blood 86:423

Davis BM, Koc ON, Lee KM, Reese JS, Schupp JE, Gerson SL (1996) Enrichment for BCNU and 0^6-benzylguanine resistant cells after transplant of G156 A MGMT transduced bone marrow progenitors in mice (meeting abstract). Blood 88:431a

Deisseroth A, Kavanagh J, Champlin R (1994) Use of safety modified retroviruses to introduce chemotherapy resistance sequences into normal hematopoietic cells for chemoprotection during the therapy of ovarian cancer: a pilot trial. Hum Gene Ther 5:1507

Delaflor-Weiss E, Richardson C, Ward M et al (1992) Transfer and expression of the human multidrug resistance gene in mouse erythroleukemia cell. Blood 80:3106–3111

Deuchars KL, Ling V (1989) Multidrug resistance in cancer chemotherapy. Semin Oncol 16:156–165

Dick JE, Magli MC, Huszar D, Phillips RA, Bernstein A (1985) Introduction of a selectable gene into primitive stem cells capable of long-term reconstitution of the hematopoietic system of W[Wv mice. Cell 42:71

Dicker AP, Volkenandt M, Schweitzer BI, Banerjee D, Bertino JR (1990) Identification and characterization of a gene from the methotrexate resistant Chinese hamster overy cell line Pro-3MTXRIII. J Biol Chem 265:8317

Dolan ME, Moschel RC, Pegg AE (1990) Depletion of mammalian 0^6-alkylguanine-DNA alkyltransferase activity by 0^6-benzylguanine provides a means to evaluate the role of this protein in protection against carcinogenenic and therapeutic alkylating agents. Proc Natl Acad Sci USA 87:5368–5372

Drach D, Zhao S, Drach J, Mahadevia R, Gattringer C, Huber H, Andreeff M (1992) Subpopulations of normal peripheral blood and bone marrow cens express a functional multidrug resistant phenotype. Blood 80:2729

Dumenco LL, Warman B, Hatzoglou M, Lim IK, Abboud SL, Gerson SL (1989) Increase in nitrosourea resistance in mammalian cells by retrovirally mediated gene transfer of bacterial 0^6-alkylguanine-DNA alkyltransferase. Cancer Res 49:6044

Dunbar CE, Cottler FM, O'Shaughnessy JA et al (1995) Retrovirally marked CD34 enriched peripheral blood and bone marrow cells constribute to long-term engraftment after autologous transplantation. Blood 85:3948

Dzierzak EA, Papayannopoulou T, Mulligan RC (1988) Lineage-specific expression of a human beta-globin gene in murine bone marrow transplant recipients reconstituted with retrovirus-transduced stem cells. Nature 331:35–41

Ellis GK, Hutchins L, Jimenez-Martin M, Pecora AL, Barnadas A, Meisenberg B, Nabholtz JM, Cortes-Funes H, Rifkin R, Chang AYC, Garrison L, George C, Giles FJ (1996) Phase III double blind randomized study of PIXY321 versus G-CSF after CEP for breast or ovarian carcinoma (meeting abstract). Blood 88:449a

Endicott JA, Ling V (1989) The biochemistry of P-glycoprotein-mediated multidrug resistance. Annu Rev Biochem 58:137

Feng M, Jackson WH Jr, Goldman CK, Rancourt C, Wang M, Dusing SK, Siegal G, Curiel DT (1997) Stable in vivo gene transduction via a novel adenoviral/retroviral chimeric vector. Nat Biotechnol 15:866–870

Flasshove M, Banerjee D, Mineishi S, Schlafstein M, Bertino JR, Moore MAS (1995) Retrovirally mediated gene transfer of a mutant human dihydrofolate reductase (DHFR) gene into progenitors from human peripheral blood (PB). Blood 85:566

Galski H, Sullivan M, Willigham MC, Chin KV, Gottesman MM, Pastan I, Merlino GT (1989) Expression of a human multidrug resistance cDNA (MDR1) in the bone marrow of transgenic mice : resistance to daunomycin-induced leukopenia. Mol Cell Biol 9:4357

Gerson SL, Miller K, Berger NA (1985) 0^6-alkylguanine-DNA alkyltransferase activity in human myeloid cells. J Chn Invest 76:2106

Gerson S, Trey J, Miller K et al (1987) Repair of 0^6-alkylguanine drug DNA synthesis in murine bone marrow hematopoietic precursors. Cancer Res 47:89–95

Gerson S, Phillips W, Kastan M, Dumenco LL, Donovan C (1996) Human CD34+ hematopoietic progenitors have low, cytokine-unresponsive 0^6-alkylguanine-DNA alkyltransferase and are sensitive to 0^6-benzylguanine plus BCNU. Blood 88:1649–1655

Gilboa E, Eglitis NU, Kantoff PW, Anderson WF (1986) Transfer and expression of cloned genes using retroviral vectors. Biotechniques 4:504

Gottesman MM, Pastan I (1988) The multidrug transporter, a double edged sword. J Biol Chem. 263:12163–12166

Greenbaum M, Letourneau S, Assar H, Schecter R, Batist G, Cournoyer D (1994) Retrovirus mediated gene transfer of rat glutathione S-transferase Yc confers alkylating drug resistance in NIH 3T3 fibroblast. Cancer Res 54:4442

Gros P, Neriah YB, Croop JM, Housman DE (1986a) Isolation and expression of a complementary DNA that confers multidrug resistance. Nature 323:728–731

Gros P, Croop J, Housman D (1986b) Mammalian multidrug resistance gene: complete cDNA sequence indicates strong homology to bacterial transport proteins. Cell 47:371

Hanania EG, Deisseroth AB (1994) Serial transplantation shows that early hematopoietic precursor cells are transduced by MDR-l retroviral vector on a mouse gene therapy model. Cancer Gene Ther 1:21

Hanenberg H, Xiao XL, Diloo D, Hashino K, Brenner MK, Kato I, Williams DA (1995) Retrovirus binds to fibronectin: implication for clinical gene therapy (meeting abstract). Stem cell gene therapy: biology and technology, Washington DC

Hesdorffer C, Antman K, Bank A et al (1994) Human MDR gene transfer in patients with advanced cancer. Hum Gene Ther 5:1151

Hesdorffer C, Ayello J, Ward M, Reiss R, Vahdat L, Fetell M, Garrett TD, Bank A, Antman K (1997) A phase I clinical trial using MDR-transduced marrow and/or peripheral blood CD34+ cells (meeting abstract). Proc ASCO 16:88a

Hickson I, Fairbarin LJ, Chinnasamy N, Dexter TM, Margison GP, Rafferty JA (1996) Protection of mammalian cells against chloroethylating agent toxicity by an 0^6-benzylguanine-resistant mutant of human 0^6-alkylguanine-DNA alkyltransferase. Gene Ther 3:868–877

Hilton J (1984) Role of aldehyde-dehydrogenase in cyclophosphamide-resistant L1210 leukemia. Cancer Res 44:5156

Hirose K, Longo D, Oppenheim JJ, Matsushima K (1993) Overexpression of mitochondrial manganese superoxide dismutase promotes the survival of tumor cells exposed to interleukin-1, tumor necrosis factor, selected anticancer drugs and ionizing radiation. FASEB J 7:361

Hockenbery DM, Oltval ZN, Yin X-M, Milliman CL, Korsmeyer SJ (1993) Bcl-2 functions in an antioxidant pathway to prevent apoptosis. Cell 75:241–251

Hoffman B, Leibermann DA (1994) Molecular controls of apoptosis; differentiation/growth arrest primary response genes, protooncogenes, and tumor suppressor genes as positive and negative modulators. Oncogene 9:1807–1812

Hofstra LS, Trope CG, Willemse PHB, Vindevoghel A, van den Bulche JM, Lahouseny M, Sklenar I, de Vries EGE (1997) Randomized trial of rhIL-3 versus placebo in prevention of bone marrow depression during first-line chemotherapy for ovarian carcinoma (meeting abstract). Proc ASCO 16:115a

Hryniuk W, Bush H (1984) The importance of dose intensity in chemotherapy of metastatic breast cancer. J Clin Oncol 2:1281

Hsu LC, Tani K, FU]iyoshi T, Kurachi K, Yoshida A (1985) Cloning of cDNAs for human aldehyde dehydrogenases 1 and 2. Proc Natl Acad Sci USA 82:3771

Hsu LC, Chang W-C, Shibuya A, Yoshida A (1992) Human stomach aldehyde dehydrogenase cDNA and genomic cloning, primary structure, and expression in Escherichia coll. J Biol Chem 267:3030

Huang R, Olsson M, Petterson U, Totterman RH (1995) Efficient adenovirus mediated gene transduction of normal and leukemic hematopoietic cells (meeting abstract). Stem cell gene therapy: biology and technology, Washington DC, p 15

Hussaon A, Lewis D, Yu M, Melera PW (1992) Construction of a dominant selectable marker using a novel dihydrofolate reductase cDNA. Gene 112:179

Isola LM, Gordon JW (1986) Systemic resistance to methotrexate in transgenic mice carrying a mutant dlhydrofolate reductase gene. Proc Natl Acad Sci USA 83:9621

Jehnek J, Kleibl K, Dexter TM, Margison GP (1988) Transfection of murine multi-potent hematopoietic stem cells with an E coli DNA alkyltransferase gene confers resistance to the toxic effects of alkylating agents. Carcinogenesis 9:81

Jehnek J, Fairbarin LJ, Dexter M, Rafferty JA, Stocking C, Ostertag W, Margison GP (1996) Long-term protection of hematopoiesis against the cytotoxic effects of multiple doses of nitrosourea by retrovirus-mediated expression of human 0^6-alkylguanine DNA-alkyltransferase. Blood 87:1957–1961

Kaina B, Fritz G, Mitra S, Coquerelle T (1991) Transfection and expression of human 0^6-methylguanine-DNA methyltransferase (MGMT) cDNA in Chinese hamster cells: the role of MGMT in protection against the genotoxic effects of alkylating agents. Carcinogenesis 12:1857

Kaleko M, Garcia JV, Osborne WR, Miller AD (1990) Expression of human adenosine deaminase in mice after transplantation of genetically modified bone marrow. Blood 75:1773

Kartner N, Ling V (1989) Multidrug resistance in cancer. Sci Am 260(3):44–51

Kastan MB, Schlaffer E, Russo JE, Colvin OM, Civin CL, Hilton J (1990) Direct demonstration of elevated aldehyde-dehydrogenase in human hematopoietic progenitor cells. Blood 75:1947

Kavanagh J, Hanania E, Giles R, Fu SO, Zu Z, Ellerson D, Wang T, Claxton D, Rahman Z, Berenson R, Heimfeld S, Cote R, Holzmayer T, Mechetner E, Dayne A, Andreeff M, Champlin R, Deisseroth AB (1996) Genetic modification of cells used for transplant following intensive therapy for ovarian cancer and breast cancer (meeting abstract 272a). Blood 88:

Kondo S, Yin D, Morimura T, Oda Y, Kikuchi H, Takeuchi J (1994) Transfection with a bcl-2 expression vector protects transplanted bone marrow from chemotherapy induced myelosuppression. Cancer Res 54:2928

Kotin R, Sinscalco M, Samulski RJ et al (1990) Site-specific integration by adeno-associated virus. Proc Natl Acad Sci USA 87:2211

Kuroda H, Sakamaki S, Kuga T, Matsunaga T, Hirayama Y, Ohi S, Takahashi Y, Niitsu Y (1996) Transfection of glutathione-s transferase p gene into CD34+ cells confers resistance to cyclophosphamide (meeting abstract 1705). Blood 88:

Laliberte J, Momparler RL (1994) Human cytidine deaminase: purification of enzyme, cloning and expression of its complementary DNA. Cancer Res 54:5401

Li MX, Banerjee D, Mineishi S, Gilboa E, Bertino JR (1994) Development of a retroviral construct containing a human mutated dihydrofolate reductase cDNA for hematopoietic stem cell transduction. Blood 83:3403

Li MX, Banerjee D, Zhao SC, et al. (1994) Development of a retroviral construct containing a human mutated dihydrofolate reductase cDNA for hematopoietic stem cell transduction. Blood 83:3403

Licht T, Aran JM, Goldenberg SK, Vieira WD, Gottesman MM, Pastan I (1995) Cytotoxic drug selection of murine bone marrow cells following transfer of the MDR1 gene increases gene expression and chemoresistance in vivo. Meeting abst. Blood 86:A960

Ludlum DB, Mehta JR, Tong WP (1986) Prevention of I-(3-deoxycytytidyl), 2-(l-deoxyguanosinyl)ethane cross-link formation in DNA by rat liver 0^6-alkylguanine-DNA alkyltransferase. Cancer Res 46:3353

Magni M, Shammah S, Schiro R, Mellado W, Dalla-Favera R, Gianni AM (1996) Induction of cyclophosphamide-resistance by aldehyde-dehydrogenase gene transfer. Blood 87:1097

Markowitz D, Goff S, Bank A (1988a) Construction and use of a safe and efficient amphotropic packaging cell line. Virology 167:400

Markowitz D, Goff S, Bank A (1988b) A safe packaging line for gene transfer: separating viral genes on two different plasmids. J Virol 62:1120

May C, Gunther R, McIver RS (1995) Protection of mice from lethal doses of methotrexate by transplantation with transgenic marrow expressing drug-resistant dihydrofolate reductase activity. Blood 86:2439–2448

McGown AT, Fox BW (1986) A proposed mechanism of resistance to cyclophosphamide and phosphoramide mustard in a Yoshida cell line in vitro. Cancer Chemother Pharmacol 17:223

McIvor RS, Simonsen CC (1991) Isolation and characterization of a variant dlhydrofolate reductase cDNA from methotrexate resistant murine L5178Y cells. Nucleic Acids Res 18:7025

McLachlin JR, Eglitis MA, Ueda K, Kantoff PW, Pastan IH, Anderson WF, Gottesman MM (1990) Expression of a human complementary DNA for the multidrug resistance gene in murine hematopoietic precursor cells with the use of retroviral gene transfer. J Natl Cancer Inst 82:1260

Mivashita T, Reed JC (1992) Bcl-2 gene transfer increases relative resistance of S49. 1 and WEHI 7. 2 lymphold cells to cell death and DNA fragmentation induced by glucocorticoids and multiple chemotherapeutic drugs. Cancer Res 52:5407–5411

Miyashita T, Reed JC (1993) Bcl-2 oncoprotein blocks chemotherapy induced apoptosis in a human leukemia cell line. Blood 81:151–157

Momparler RL, Ehopoulos N, Bovenzi V, Letourneau S, Greenbaum M, Cournoyer D (1996) Resistance to cytosine arabinoside by retrovirany mediated gene transfer of human cytidine deaminase into murine fibroblast and hematopoletic cells. Cancer Gene Ther 3:331

Moore KA, Deisseroth AB, Reading CL, Williams DE, Belmont JW (1992) Stromal support enhances cell-free retroviral vector transduction of human bone marrow long-term culture-initiating cells. Blood 79:1393

Moore MAS (1995) Hematopoietic reconstruction: New approaches. Clin Cancer Res 1:3–9

Moore MAS, Hoskins I (1994) Ex vivo expansion of cord blood-derived stem cells and progenitors. Blood cells 20:468

Moore MAS, Leonard JP, Flasshove M, Bertino J, Gallardo H, Sadelain M (1996) Review: gene therapy – the challenge for the future. Ann Oncol 7:53

Moreb J, Schweder M, Suresh A, Zucah J (1996) Overexpression of the human aldehyde dehydrgenase class I results in increased resistance to 4-hydroperoxycyclophosphamide. Cancer Gene Ther 3:24

Moritz T, Mackay W, Glassner BJ, Williams DA, Samson L (1995) Retrovirus mediated expression of a DNA repair protein in bone marrow protects hematopoietic cells from nitrosourea-induced toxicity in vitro and in vivo. Cancer Res 55:2608

Nakahara S, Takebe N, Mineishi S, Zhao SC, Erickan-Abali E, Banerjee D, Sadelain M, Moore MAS, Bertino JR (1997) MFG-based retroviral vectors carrying human DHFR mutants confer higher resistance to MTX than N2-based double copy vectors in hematopoietic cells. meeting abstract. Proc Am Assoc Cancer Res 38:384

Naldini L, Blomer U, Gallay P, Ory D, Mulligan R, Gage FH, Verma I, Trono D (1996) In vivo gene delivery and stable transduction of nondividing cells by a lentiviral vector. Science 272:263

Naumovski L, Cleary ML (1994) Bcl-2 inhibits apoptosis associated with terminal differentiation of HL-60 myelold leukemia cells. Blood 83:2261–2267

Neff T, Blau A (1996) Forced expression of cytidine deaminase confers resistance to cytosine arabinoside and gemcitabine. Exp Hematol 24:1340

Nolta JA, Seader LS, Barranger JA, Kohn DB (1990) Expression of human glucocerebrosidase in murine long-term bone marrow cultures after retroviral vector-mediated transfer. Blood 75:787

Ogawa M (1989) Effects of hematopoietic growth factors on stem cells in vitro. Hematol Oncol Clin North Am 3:453

Oltvai ZN, MilliLman CL, Korsmeyer SJ (1993) Bcl-2 heterodimerizes in vivo with a conserved homolog Bax, that accelerates programmed cell death. Cell 74:609–619

O'Shaughessy J, Cowan K, Nienhuis A (1994) Retroviral mediated transfer of the human multidrug resistance gene (MDR-1) into hematopoietic stem cells during autologous transplantation after intensive chemotherapy for metastatic breast cancer. Hum Gene Ther 5:891

Pastan I, Gottesman MM (1987) Multidrug resistance in human cancer. N Engl J Med 316:1388–1393

Pastan I, Gottesman MM (1991) Multidrug resistance. Annu Rev Med 42:277

Pegg A (1984) Methylation of the 06-position of guanine in DNA is the most likely initiating event in carcinogenesis by methylating agents. Cancer Invest 2:223–231

Pegg A (1990) Mammalian 0^6-alkylguanine-DNA alkyltransferase. Regulation and importance in response to alkylating carcinogenic and therapeutic agents. Cancer Res 50:6119–6129

Pegg AE, Weist L, Foote R et al (1983) Purification and properties of 0^6-methylguanine-DNA methyltransferase from rat liver. J Biol Chem 258:2327–2333

Pizzorno G, Chang YM, McGuire JJ, Bertino JR (1989) Inherent resistance of human squamous carcinoma cell lines to MTX as a result of decreased polyglutamylation of this drug Cancer Res 49:5275

Podda S, Ward M, Himelstein A, Richardson C, de la Flor-Weiss E, Smith L, Gottesman M, Pastan I, Bank A (1992) Transfer and expression of the human multiple drug resistance gene into live mice. Proc Natl Acad Sci USA 89:9676

Reece DE, Connors JM, Spinelli JJ, Barnett MJ, Fairey RN, Klingemann HG, Nantel SH, O'Reilly S, Shepherd JD, Sutherland HJ et al (1994) Intensive therapy with cyclophosphamide, carmustine, etoposide+/- cisplatin, and autologous bone marrow transplantation for Hodgkin's disease in first relapse after combination chemotherapy. Blood 83:1193

Reed JC, Tsujimoto Y, Epstein SF, Cuddy M, Slablak T, Nowell PC, Croce CM (1989) Regulation of bcl-2 gene expression in lymphoid cell lines containing normal no 18 or t(14: 18) chromosomes. Oncogene Res 4:271–282

Rekha GK, Sreerama L, Sladek NE (1994) Intrinsic cellular resistance to oxazaphosphorines exhibited by a human colon carcinoma cell line expressing relatively large amounts of a class-3 aldehyde dehydrogenase. Biochem Pharmacol 48:1943

Richardson C, Bank A (1995) Preselection of transduced murine hematopoietic stem cell populations leads to increased long-term stability and expression of the human multiple drug resistance gene. Blood 86:2579

Roninson IB, Chin JE, Choi KG, Gros P, Housman DE, Fojo A, Shen DW, Gottesman MM, Pastan I (1986) Isolation of human mdr DNA sequences amplified in multidrug-resistant KB carcinoma cells. Proc Natl Acad Sci USA 83:4538

Rosti G, Albertazzi L, Salvioni R, Pizzocaro G, Cetto GL, Bassetto NU, Marangolo M (1992) High dose chemotherapy supported with autologous bone marrow transplantation (ABMT) in germ cell tumors: a phase two study. Ann Oncol 3:809

Russo JE, Hilton J (1988) Characterization of cytosolic aldehyde-dehydrogenase from cyclophosphamide resistant L1210 cells. Cancer Res 48:2963

Schweitzer BI, Srimatkadada S, Gritsman H, Sheridan R, Venkataraghavan R, Bertino JR (1989) Probing the role of two hydrophobic active site residues in dihydrofolate reductase by site directed mutagenesis. J Biol Chem 264:20786

Sha AJ, Smogorzewska EM, Hannum C et al (1995) FLT3 ligand(FL) is a potent inducer of proliferation of highly quiescent human bone marrow CD34+ CD38- cells. Blood 86:423 (abstract)

Simonsen CC, Levinsos AD (1983) Isolation and expression of an altered mouse dihydrofolate reductase cDNA. Proc Natl Acad Sci USA 80:2495

Sorrentino BP, Brandt SJ, Bodine D, Gottesman M, Pastan I, Cline A, Nienhuis AW (1992) Selection of drug-resistant bone marrow cells in vivo after retroviral transfer of human MDR1. Science 257:99

Sorrentino BP, McDonagh KT, Woods D et al (1995) Expression of retroviral vectors containing the human multidrug resistance 1 cDNA in hematopoietic cells of transplanted mice. Blood 86:491–501

Spencer HT, Sleep SEH, Rehg JE, Blakley RL, Sorrentino BP (1996) A gene transfer strategy for making bone marrow cells resistant to trimetrexate. Blood 87:2579

Sreerama L, Sladek NE (1994) Identification of a methylcholanthrene-induced aldehyde dehydrogenase in a human breast adenocarcinoma cell line exhibiting oxazaphosphorine-specific aquired resistance. Cancer Res 54:2176

Stiff PJ, Oldenberg D, Hsi E, Chen B, Douvile J, Burhop S, Bayer R, Peace D, Malhotra D, Kerger C, Armstrong D, Muller T (1997) Successful hematopoietic engraftment following high-dose chemotherapy using only ex vivo expanded bone marrow grown in Aastrom (stromal-based) bioreactors. meeting abstract. Proc Am Soc Clin Oncol 16:88a

Tano K, Shiota S, Collier J, Foote RS, Mitra S (1990) Isolation and structural characterization of a CDNA clone encoding the human DNA repair protein for 0^6-alkylguanine. Proc Natl Acad Sci USA 87:686–690

Tew KD (1994) Glutathione associated enzymes in anticancer drug resistance. Cancer Res 54:4313

Tong W, Kirk M, Ludlum D (1982) Formation of the cross-link 1-[N3-deoxycyctidyl], 2-[NI-deoxyguanosinyl]-ethane in DNA treated with N, NI-bis(2-chloroethyl)-N-nitrosourea. Cancer Res 42:3102–3105

Tsujimoto Y, Cossman J, Jaffe B, Croce CM (1985) Involvement of the bcl-2 gene in human follicular lymphoma. Science 228:40

Veeis DJ, Sorenson CM, Shutter JR, Korsmeyer SJ (1993) Bcl-2 deficient mice demonstrate fulminant lymphoid apoptosis, polycystic kidneys, and hypopigmented hair. Cell 75:229

von Eitzen U, Meier-Tackmann D, Agarwal DP, Goedde HW (1994) Detoxification of cyclophosphamide by human aldehyde dehydrogenase isozymes. Cancer Lett 76:45

Walker S, Sofia M, Kakarla R, Kogan N, Wierichs L, Longley CB, Bruker K, Axelrod HR, Midha S, Babu S, Kahne D (1996) Cationic facial amphiphiles: a promising class of transfection agents. Proc Natl Acad Sci USA 93:1585

Ward M, Richardson C, Pioll P et al (1994) Transfer and expression of the human multiple drug resistance gene in human CD34+ cells. Blood 84:1408–1414

Williams DA, Hsieh K, DeSilva A, Mulligan RC (1987) Protection of bone marrow transplant recipients from lethal doses methotrexate by the generation of methotrexate resistant bone marrow. J Exp Med 166:21

Wilson JM, Danos O, Grossman M, Raulet DH, Mulligan RC (1990) Expression of human adenosine deaminase in mice reconstituted with retrovirus-transduced hematopoietic stem cells. Proc Natl Acad Sci USA 87:439

Wong GHW, Elwell JH, Oberly LW, Goeddel DV (1989) Manganous superoxide dismutase is essential for cellular resistance to cytotoxicity of tumor necrosis factor. Cell 58:923

Yoshida A, Dave V, Han H, Scalon KJ (1993) Enhanced transcription of the cytosolic ALDH gene in cyclophosphamide resistant human carcinoma cells. Enzymol Mol Biol Carb Metab 4:63

Zhao SC, Li MX, Banerjee D, Mineishi S, Gilboa E, Bertino JR (1994) Long-term protection of recipient mice from lethal doses of methotrexate by marrow infected with a double copy vector retrovirus confining a mutant dihydrofolate reductase. Cancer Gene Ther 1:27

Zhao SC, Banerjee D, Mineishi S, Bertino JR (1996) Mice bearing a transplanted mammary carcinoma tolerate curative doses of methotrexate (MTX) after transplantation with a retroviral construct containing a mutated dihydrofolate reductase (MDHFR) cDNA. AACR abstract

Zheng CF, Wang TTY, Weiner H (1993) Cloning and expression of the full-length cDNA's encoding human liver class 1 and class 2 aldehyde dehydrogenase. Alcohol Clin Exp Res 17:828–831

9 Pro-drug Gene Therapy for Prostate Cancer

H.L. Adler and P.T. Scardino

9.1 Introduction

Despite a declining incidence of new cases, prostate cancer remains the most common solid organ malignancy in American men (Landis et al. 1998). In 1998, approximately 184 500 new cases of prostate cancer are predicted, accounting for 29% of all male cancers. Moreover, prostate cancer will account for approximately 39 200 cancer deaths in 1998. This will comprise 13% of all cancer deaths in men, second only to lung cancer (Landis et al. 1998). Curative measures for prostate cancer include radical prostatectomy and radiation therapy; however, patients with locoregional and/or distant spread of disease cannot be cured. Following a radical prostatectomy, patients with positive surgical margins, seminal vesicle involvement and/or positive lymph nodes are at increased risk for disease recurrence and shortened survival (Rosen et al. 1992). From 13% to 70% of patients who undergo a radical prostatectomy are clinically understaged and are postoperatively found to have tumor extension outside the prostate (Ohori et al. 1995; Paulson 1994; Trapasso et al. 1994; Walsh et al. 1994; Zeitman et al. 1994; Zincke et al. 1994). Patients with metastatic prostate cancer are incurable. Androgen deprivation has been the standard treatment for these patients for over 50 years; however, survival ranges from 2 to3 years once metastases are identified and patients eventually develop hormone-refractory disease (Huggins and Hodges 1941; Crawford et al. 1990). As a result of the inherent failure rates in the treatment of clinically localized disease, combined with the inability to cure metastatic disease, gene therapy has been investigated in the laboratory (and now in the clinic) as a potential new adjuvant treatment for prostate cancer. It may eventually become sufficiently effective to use alone to control this disease.

9.2 Types of Cancer Gene Therapy

Gene therapy for cancer can be divided into several categories (Table 1; Anderson 1994). Transfer of suppressor genes including p53 in experimental models has been shown to be effective in suppressing the growth of lung tumors, head and neck squamous cell carcinomas, and human prostate cancer in a subcutaneous model (Fujiwara et al. 1994; Clayman et al. 1995; Asgari et al. 1997). In contrast, transfer of p21 in an orthotopic mouse model of prostate cancer was significantly more effective than p53 in suppressing tumor growth and extending animal survival (Eastham et al. 1996). Insertion of antisense c-myc also showed

Table 1. Categories of gene therapy being investigated for the management of malignancies

Suicide gene therapy	HSV-tk; cytosine deaminase
Replacement of tumor suppressor gene and/or anti-oncogenes	p53; antisense k-ras
Replacement of HLA genes	HLA-B7 into B7 deficient melanomas
Vaccine therapy	Insertion of cytokine genes into irradiated tumor cells in vitro
Multidrug resistance genes to protect bone marrow	

From Anderson 1994.

promise as a potential therapy for prostate cancer (Balaji et al. 1997). Tumor vaccines with cells engineered to secrete various cytokines including interleukin-2, granulocyte-macrophage colony stimulating factor, and interferon-γ have been evaluated in animal models of prostate cancer with encouraging results. Better results, however, were observed in subcutaneous tumors than in orthotopic tumors, which has important implications for the treatment of human prostate cancer (Vieweg et al. 1994; Moody et al. 1994).

9.3 Pro-drug Gene Therapy

Currently, the most promising form of gene therapy for the management of prostate cancer is via the transduction of genes that activate pro-drugs to produce cytotoxicity. One specific type of cytotoxic (or suicide) gene therapy is viral vector transduction of the herpes simplex virus-thymidine kinase (HSV-tk) gene followed the systemic administration of ganciclovir (Moolten 1986). Ganciclovir is phosphorylated by the HSV-tk protein to a monophosphate and subsequently triphosphorylated by mammalian kinases. The ganciclovir is thereby converted to its active metabolite and then leads to direct cell death of actively dividing cells by termination of DNA synthesis (Fyfe et al. 1978; Elion 1983; Oliver et al. 1985). In addition to direct cytotoxicity, the combination of HSV-tk and ganciclovir has been shown have significant toxicity on neighboring neoplastic cells. This phenomenon is known as the "bystander effect" and can lead to the death of over 90% of cells when only 30% of the cancer cells have been transduced with HSV-tk (Freeman et al. 1993). The bystander effect may be due to several mechanisms including the

Table 2. Mechanisms behind the bystander effect

In vitro	Gap junctions/connexins; uptake of apoptotic bodies; passage of toxic metabolites
In vivo	Stimulation of the immune response: increase in TNF, IL-6, IL-1; up-regulation of B7 and ICAM Inhibition of endothelial cells

TNF, tumor necrosis factor; IL, interleukin.

From Freeman et al. 1993; Elshami et al. 1996; Vile et al. 1994; Hamel et al. 1996; Ramesh et al. 1996a,b.

presence of gap junctions, uptake of apoptotic bodies, and stimulation of the immune system (Table 2; Freeman et al. 1993; Elshami et al. 1996; Vile et al. 1994; Hamel et al. 1996; Ramesh et al. 1996a,b).

9.3.1 Laboratory Models

The efficacy of adenoviral-mediated HSV-tk+ganciclovir therapy has been evaluated in many tumor models, including melanoma, colon cancer, glioma, squamous cell carcinoma of the head and neck, and mesothelioma (Bonnekoh et al. 1995; Chen et al. 1994, 1995; Perez-Cruet et al. 1994; Ram et al. 1994; O'Malley et al. 1995; Elshami et al. 1996). In each instance, the investigators demonstrated tumor growth suppression, prolonged survival, and even cure of disease in the treated animals. Several years ago, we initiated in vitro as well as in vivo experiments with adenoviral-mediated HSV-tk (ADV/HSV-tk) and ganciclovir gene therapy in order to assess efficacy in the treatment of prostate cancer.

In 1996, Eastham et al. reported on our experience with ADV/HSV-tk+ganciclovir in both in vitro and in vivo model systems. As shown in Fig. 1, an inhibitory effect on the proliferation of DU 145, PC3, and LNCaP cell lines in vitro was observed. A similar effect was observed in vitro with the RM-1 cell line (a primary mouse prostatic tumor derived from the mouse prostate reconstitution model system; Thompson et al. 1989). In this same report, subcutaneous prostate tumors were established in syngeneic C57/BL6 mice and then divided into treatment and control groups. Animals treated with ADV/HSV-tk+ganciclovir demon-

Fig. 1A–E. In vitro sensitivity of mouse and human prostate cancer cell lines to herpes simplex virus-thymidine kinase (HSV-tk)+ganciclovir (GCV). Cell lines were infected at the indicated multiplicity of infection (MOI) with adenovirus (ADV)/HSV-tk. After 24 h the media was changed and cells treated with either 10 µg/ml ganciclovir (*closed box*) or phosphate-buffered saline (PBS) (*open box*), both in serum-free medium. After 72 h the number of surviving cells was determined with a Coulter counter. **A** murine RM1 cells; **B** PC-3 cells, **C** DU145 cells; LNCaP cells grown in the presence **D** or absence **E** of 10 nM testosterone

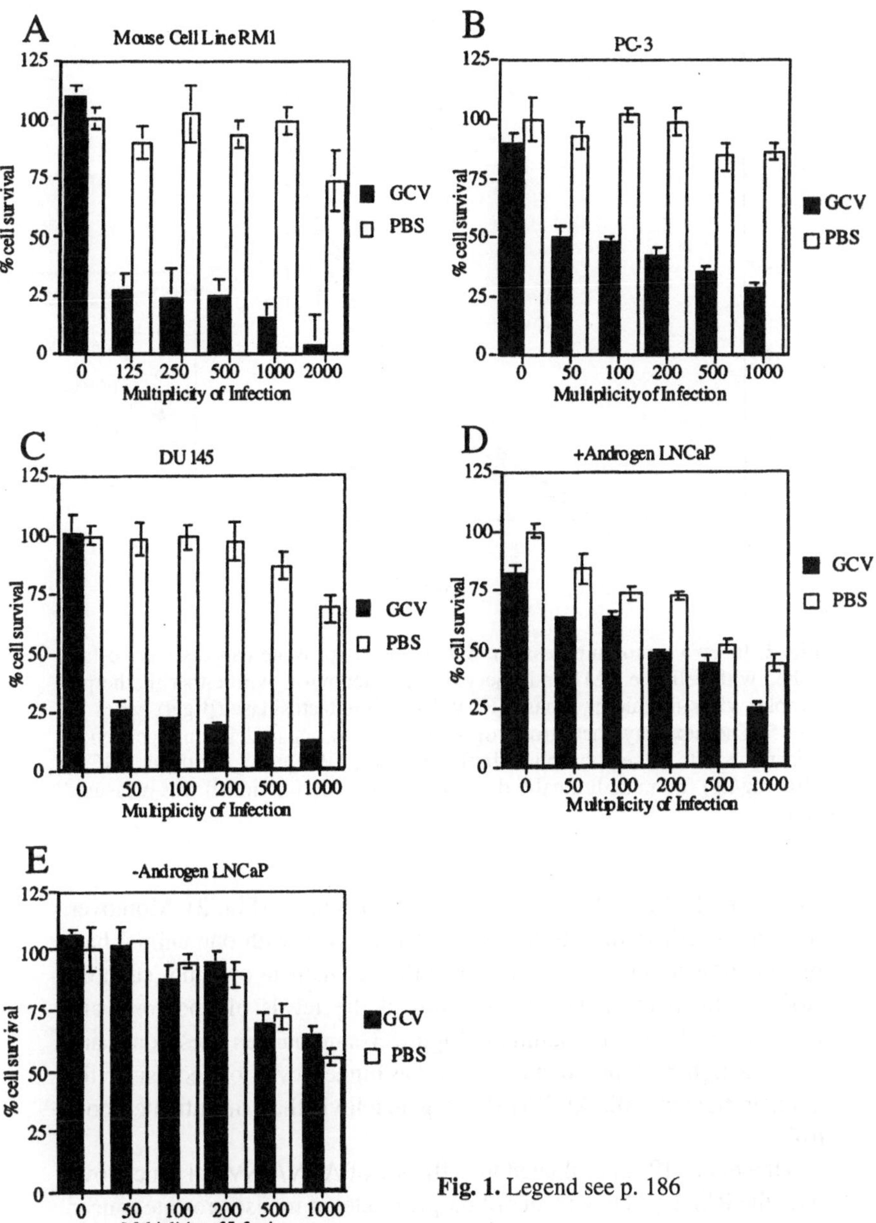

Fig. 1. Legend see p. 186

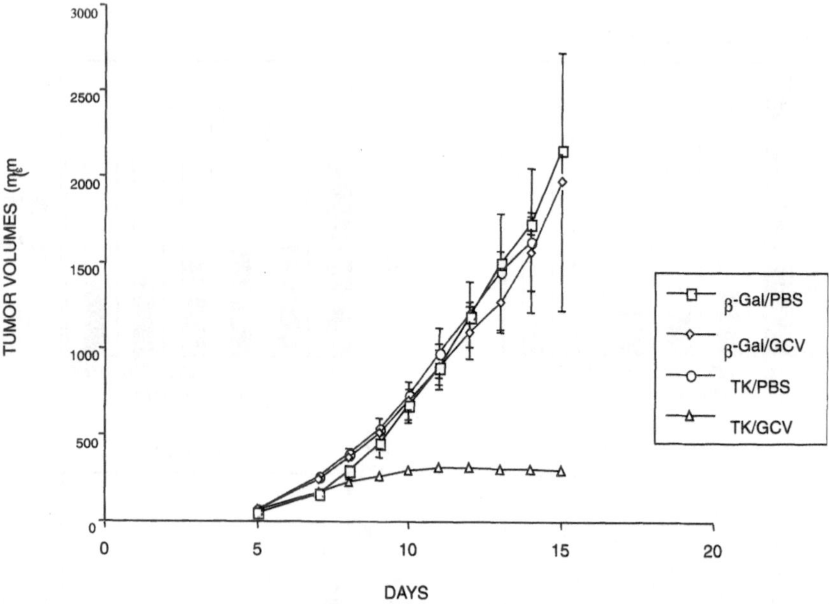

Fig. 2. In vivo tumor suppression. Subcutaneous prostate cancers were established with cell line RM1 and injected with adenovirus with either the herpes simplex virus-thymidine kinase (HSV-tk) or β-galactosidase (β-gal) genes on day 5. The next day each group of animals (n=5) began daily intraperitoneal injections with either phosphate-buffered saline (PBS) or ganciclovir (GCV) (100 mg/kg) for an additional 6 days. The tumor volume (mm^3) was measured daily

strated marked growth suppression of their tumors (Fig. 2). Moreover, treated animals demonstrated increased survival, with one animal having a complete response. To evaluate the mechanism of tumor suppression, Eastham et al. (1996) determined the levels of apoptosis and necrosis in the treated animals. Figure 3 summarizes these data and shows a higher apoptotic index as well as higher levels of necrosis in the animals treated with ADV/HSV-tk+ganciclovir than in untreated controls.

Hall et al. (1997) evaluated the efficacy of ADV/HSV-tk+ganciclovir with the RM-1 cell line in an orthotopic model of mouse prostate cancer. In addition to demonstrating growth suppression of tumor and increased

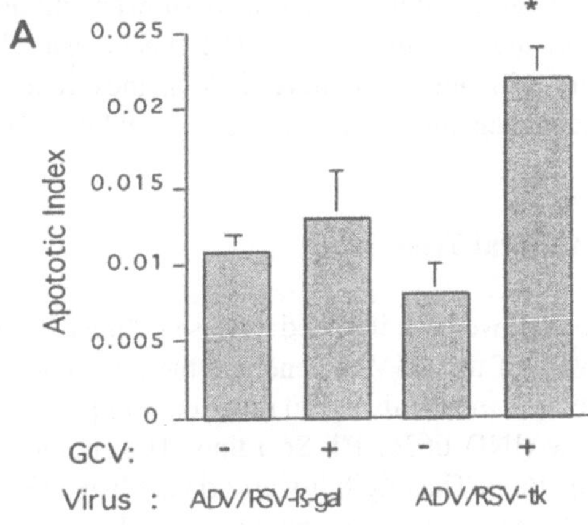

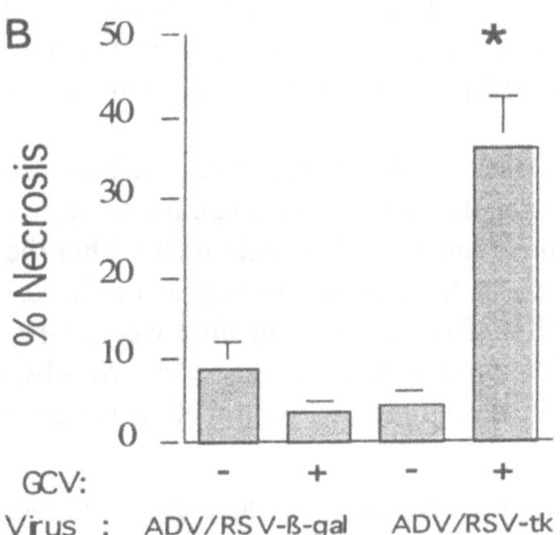

Fig. 3A, B. Apoptosis and necrosis in adenovirus (ADV)/herpes simplex virus-thymidine kinase (HSV-tk) plus ganciclovir (GCV)-treated tumors. **A** apoptotic index (AI) expressed as number of apoptotic cells per 100 cells in viable portions of tumors injected with the indicated vector and treated with either phosphate-buffered saline (PBS) or ganciclovir (GCV). **B** Extent of necrosis expressed as percentage of the total tumor area in four to six sections from tumors injected with the indicated vector and treated with either PBS or GCV

survival in the treated animals, they also showed inhibition of both spontaneous and induced metastases. The mechanisms behind these observations are still unknown; nevertheless, they represent exciting prospects in the management of prostate cancer in human clinical trials.

9.3.2 Phase I Clinical Trial

Based on these data, we have initiated a Phase I clinical trial of adenoviral vector delivery of the HSV-tk gene and the intravenous administration of ganciclovir in men with a local recurrence of prostate cancer after radiation therapy (IND 6636, PT Scardino, TC Thompson, and SLC Woo co-investigators). The eligibility criteria are listed in Table 3. Once enrolled, patients receive a transrectal ultrasonically guided injection of the ADV vector. The next morning patients receive intravenous ganciclovir 5 mg/kg every 12 h for 14 days. The patients are followed with serial physical examinations, prostate specific antigen (PSA) levels, blood counts, chemistries, cultures, transrectal ultrasounds, and prostate biopsies.

We treated a series of patients in a dose escalation trial to establish the maximum tolerated concentration when the vector was delivered in a constant volume along a single needle tract within the prostate. We found no evidence of toxicity or efficacy at the lowest dose, 1×10^8 infectious units (IU). Consequently, the dose was escalated logarithmically in groups of three to five patients each. We observed minimal toxicity with mild elevations in liver function parameters in several

Table 3. Eligibility criteria for entry into Phase I trial of adenoviral-mediated HSV-tk+ganciclovir gene therapy

Biopsy proven prostate cancer at least 1 year after definitive radiation therapy
Rising prostate-specific antigen (PSA) level on three separate occasions at least 2 weeks apart (but less than 20 ng/ml)
No evidence of metastatic disease: CT scan, bone scan
No prior hormonal therapy
Normal baseline organ function (determined by complete blood count, serum chemistries, and liver function tests)
No evidence of acute infection (bacterial, viral, and/or fungal)
Negative HIV status

patients. At the highest dose level, one patient exhibited thrombocytopenia requiring platelet transfusions and evidence of hepatic injury based on serum liver function tests. These abnormalities resolved after ganciclovir therapy was discontinued. The mechanism(s) behind these abnormalities are unclear at this time.

Remarkably, there was evidence of efficacy in this Phase I trial with several patients exhibiting a decline in the serum PSA level greater than 50% lasting for several months. In other patients, PSA levels rose post-treatment, but subsequently stabilized. This study suggests that intraorgan delivery of ADV-mediated HSV-tk+ganciclovir gene therapy for clinically localized prostate cancer is safe and may be active against this disease. Further analysis is necessary to determine the systemic, cellular, and molecular changes stimulated by such suicide gene therapy.

9.4 Conclusion

In conclusion, pro-drug gene therapy with HSV-tk+ganciclovir has significant potential in the management of prostate cancer patients. Our clinical data show that HSV-tk+ganciclovir gene therapy is safe. We will continue to follow the responding patients to determine the durability of their responses. Further studies have been initiated to determine the effects of repeated courses of this form of therapy, the optimal number and distribution of sites of injection within the prostate, and the possible anti-metastatic effect of local administration of gene therapy. We plan to explore the efficacy of HSV-tk+ganciclovir gene therapy in combination with definitive local therapy such as radical prostatectomy and external beam irradiation therapy. We believe, however, that durable efficacy will require a combination of cytotoxic gene therapy with cytokine vectors and other techniques to enhance the host response. Improved vector systems are needed to provide greater transduction and less risk of immune response to the vector itself.

Acknowledgements. Supported in part by a Specialized Program of Research Excellence (SPORE) in prostate cancer grant (CA58204) from the National Cancer Institute.

References

Anderson FW (1994) Gene therapy for cancer. Hum Gene Ther 5:1–2

Asgari K, Sesterhenn IA, McLeod DG, Cowan K, Moul JW, Seth P, Srivastava S (1997) Inhibition of the growth of pre-established subcutaneous tumor nodules of human prostate cancer cells by single injection of the recombinant adenovirus p53 expression vector. Int J Cancer 71:377–382

Balaji KC, Koul H, Mitra S, Maramag C, Reddy P, Menon M, Malhotra RK, Laxmann S (1997) Antiproliferative effects of c-myc antisense oligonucleotide in prostate cancer cells: a novel therapy in prostate cancer. Urology 50:1007–1015

Bonnekoh B, Greenhalgh DA, Bundman DS, Eckhardt JN, Longley MA, Chen S-H, Woo SLC, Roop DR (1995) Inhibition of melanoma growth by adenoviral-mediated HSV thymidine kinase gene transfer in vivo. J Invest Dermatol 104:313–317

Chen S-H, Chen XHL, Wang Y, Kosai K-I, Finegold MJ, Rich SS, Woo SLC (1995) Combination gene therapy for liver metastasis of colon carcinoma in vivo. Proc Natl Acad Sci USA 92:2577–2581

Chen S-H, Shine HD, Goodman JC, Grossman RG, Woo SLC (1994) Gene therapy for brain tumors: regression of experimental gliomas by adenovirus-mediated gene transfer in vivo. Proc Natl Acad Sci USA 91:3054–3057

Clayman GL, El-Naggar AK, Roth JA, Zhang W-W, Goepfert H, Taylor DL, Liu T-J (1995) In vivo molecular therapy with p53 adenovirus for microscopic residual head and neck squamous carcinoma. Cancer Res 55:1–6

Crawford ED, Blumenstein BA, Goodman PJ, Davis MA, Eisenberger MA, McLeod DG, Spaulding JT, Benson RB, Dorr FA (1990) Leuprolide with and without flutamide in advanced prostate cancer. Cancer 66 [Suppl]:1039–1044

Eastham JA, Chen S-H, Sehgal I, Yang G, Timme TL, Hall SJ, Woo SLC, Thompson TC (1996) Prostate cancer gene therapy: herpes simplex virus thymidine kinase gene transduction followed by ganciclovir in mouse and human prostate cancer models. Hum Gene Ther 7:515–523

Eastham JA, Hall SJ, Sehgal I, Wang J, Timme TL, Yang G, Connell-Crowley L, Elledge SJ, Zhang W-W, Harper JW, Thompson TC (1996) In vivo gene therapy with p53 or p21 adenovirus for prostate cancer. Cancer Res 55:5151–5155

Elion GB (1983) The biochemistry and mechanism of action of acyclovir. J Antimicrob Chemother 12:9–17

Elshami AA, Kucharczuk JC, Zhang HB, Smythe WR, Hwang HC, Litzky LA, Kaiser LR, Alebelda SM (1996) Treatment of pleural mesothelioma in an immunocompetent rat model utilizing adenoviral transfer of the herpes simplex virus thymidine kinase gene. Hum Gene Ther 7:141–148

Elshami AA, Saavedra A, Zhang, H Kucharczuk JC, Spray DC, Fishman GI, Amin KM, Kaiser LR, Albelda SM (1996) Gap junctions play a role in the 'bystander effect' of the herpes simplex virus thymidine kinase/ganciclovir system in vitro. Gene Ther 3:85–92

Freeman SM, Abboud CN, Whartenby KA, Packman CH, Koeplin DS, Moolten FL (1993) The bystander effect: tumor regression when a fraction of the tumor mass is genetically modified. Cancer Res 53:5274–5283

Freeman SM, Abboud CN, Whartenby KA, Packman CH, Koeplin DS, Moolten FL, Abraham GN (1993) The "bystander effect": tumor regression when a fraction of the tumor mass is genetically modified. Cancer Res 53:5273–5283

Fujiwara T, Cai DW, Georges RN, Mukhopadhyay T, Grimm EA, Roth JA (1994) Therapeutic effect of a retroviral wild-type p53 expression vector in an orthotopic lung cancer model. J Natl Cancer Inst 86:1458–1462

Fyfe JA, Keller PM, Furman PA, Miller RL, Elion GB (1978) Thymidine kinase from herpes simplex virus phosphorylates the new antiviral compound, 9-(2- hydroxyethoxymethyl) guanine. J Biol Chem 253:8721–8727

Hall SJ, Mutchnik SE, Chen S-H, Woo SLC, Thompson TC (1997) Adenovirus mediated herpes simplex virus thymidine kinase gene and ganciclovir therapy leads to systemic activity against spontaneous and induced metastasis in an orthotopic mouse model of prostate cancer. Int J Cancer 70:183–187

Hamel W, Magnelli L, Chiarugi VP, Israel MA (1996) Herpes simplex virus thymidine kinase/ganciclovir – mediated apoptotic death of bystander cells. Cancer Res 56:2697–2702

Huggins C, Hodges C (1941) Studies on prostate cancer. I. The effect of castration, of estrogen and of androgen injection on serum phosphatases in metastatic carcinoma of the prostate. Cancer Res 1:293–297

Landis SH, Murray T, Bolden S, Wingo PA (1998) Cancer Stat CA 48:6–29

Moody DB, Robinson JC, Ewing CM, Lazenby AJ, Isaacs WB (1994) Interleukin-2 transfected prostate cancer cells generate a local anitumor effect in vivo. Prostate 24:244–251

Moolten FL (1986) Tumor chemosensitivity conferred by inserted herpes thymidine kinase genes: paradigm for a prospective cancer control study. Cancer Res 46:5276–5281

O'Malley BW Jr, Chen S-H, Schwartz MR, Woo SLC (1995) Adenovirus-mediated gene therapy for human head and neck squamous cell cancer in a nude mouse model. Cancer Res 55:1080–1085

Ohori M, Wheeler TM, Kattan MW, Goto Y, Scardino PT (1995) Prognostic significance of positive surgical margins in radical prostatectomy specimens. J Urol 154:1818–1824

Oliver S, Bubley G, Crumpacker C (1985) Inhibition of HSV-transformed murine cells by nucleoside analogs 2'-NDG and 2'-nor-GMP. Virology 145:84–93

Paulson DF (1994) Impact of radical prostatectomy in the management of clinically localized disease. J Urol 152:1826–1830

Perez-Cruet MJ, Trask TW, Chen S-H, Goodman JC, Woo SLC, Grossman RG, Shine HD (1994) Adenovirus-mediated gene therapy of experimental gliomas. J Neurosci Res 39:506–511

Ram Z, Walbridge S, Shawker T, Culver KW, Baese RM, Oldfield EH (1994) The effect of thymidine kinase transduction and ganciclovir therapy on tumor vasculature and growth of 9L gliomas in rats. J Neurosurg 81:256–260

Ramesh R, Marroji AJ, Munshi A, Abboud C, Freeman SM (1996) Expression of costimulatory molecules: B7 and ICAM up-regulation after treatment with a suicide gene. Cancer Gene Ther 3:373–384

Ramesh R, Marroji AJ, Munshi A, Abboud C, Freeman SM (1996) In vivo analysis of the 'bystander effect': a cytokine cascade. Exp Hematol 24:829–838

Rosen MA, Goldstone L, Lapin S, Wheeler T, Scardino PT (1992) Frequency and location of extracapsular extension and positive surgical margins in radical prostatectomy specimens. J Urol 148:331

Thompson TC, Southgate J, Kitchener G, Land H (1989) Multi-stage carcinogenesis induced by ras and myc oncogenes in a reconstituted model. Cell 56:917–930

Trapasso JG, DeKernion JB, Smith RB, Dorey F (1994) Incidence and significance of detectable levels of serum prostate specific antigen after radical prostatectomy. J Urol 152:1821–1825

Vieweg J, Rosenthal FM, Bannerji R, Heston WDW, Fair WR, Gansbacher B, Gilboa E (1994) Immunotherapy of prostate cancer in the Dunning rat model: use of cytokine gene modified tumor vaccines. Cancer Res 54:1760–1765

Vile RG, Nelson JA, Castleden S, Chong H, Hart IR (1994) Systemic gene therapy of murine melanoma using tissue specific expression of the HSVtk gene involves an immune component. Cancer Res 54:6228–6234

Walsh PC, Partin AW, Epstein JI (1994) Cancer control and quality of life following retropubic prostatectomy: results at 10 years. J Urol 152:1831–1836

Zeitman AL, Edelstein RA, Coen JJ, Babayan RK, Krane RJ (1994) Radical prostatectomy for adenocarcinoma of the prostate: the influence of preoperative and pathologic findings on biochemical disease-free outcome. Urology 43:828–833

Zincke H, Oesterling JE, Blute ML, Bergstralh EH, Myers RP, Barrett DM (1994) Long-term (15 years) results after radical prostatectomy for clinically localized (Stage T2c or lower) prostate cancer. J Urol 152:1850–1857

10 Hematopoietic Ex Vivo Gene Transfer

H. Glimm, C. von Kalle, R. Henschler, and R. Mertelsmann

10.1 Introduction

The goal of somatic gene therapy is to therapeutically benefit an individual by transferring genetic information to non-germline cells. Genes can be transferred either to replace the function of a defective gene in order to add genetic information and thereby provide an additional function to a target cell or to suppress the function of an already active gene in the target cell population.

Hematopoietic stem cells are ideal targets for somatic gene therapy because of their longevity, their unique potential of self-renewal ,and their ability to differentiate along each of the hematopoietic lineages. Gene therapy by successful engraftment of pluripotent hematopoietic stem cells after ex vivo gene transfer would lead to hematopoietic repopulation of the recipient with a high number of genetically modified cells. Cells containing the transferred gene would be present for an extended period of time, possibly for the entire lifespan of the treated individual.

The potential applications of hematopoietic gene therapy include a broad variety of congenital and acquired diseases. Patients with rare, inherited single gene defects of the hematopoietic system (e.g., adenosine deaminase deficiency, leukocyte adherence deficiency, chronic granulomatous disease, sphingolipid and other macrophage storage disorders, thalassemia, sickle cell anemia) as well as non-hematopoietic dysfunctions (hemophilia A, B) would benefit from efficient genetic modification of the hematopoietic system. New concepts for the treatment of much more frequently acquired genetic diseases such as HIV and cancer are under investigation.

10.2 Vector Systems

Gene transfer can be accomplished by a broad variety of viral or non-viral vector systems, including the direct uptake of naked DNA, receptor-mediated uptake of protein-coated DNA in the form of genetically modified or synthetic viruses, and the fusion of subcellular compartments containing complete or partial chromosomes. Key differences between the various modalities for hematopoietic gene transfer relate to transduction efficiency and the potential for stable integration of the transgene into the target cell genome, a prerequisite for gene transfer into multiple generations of differentiated blood cells.Therefore, only efficiently integrating viral vector systems derived from the adeno-associated virus (AAV) or from retroviruses are currently considered relevant for stem cell gene therapy.

Most available studies on gene transfer into hematopoietic stem cells have been performed with murine leukemia virus (MLV)-derived vector systems because of their high transduction efficiency, so far unsurpassed biosafety and their inherent genetic stability (Miller and Rosman 1989; Miller et al. 1991; Nienhuis 1994). A disadvantage of retroviral vectors derived from MLV is the requirement of mitosis for their entry into the nucleus and efficient integration. No other transport mechanism seems to permit entry of the viral preintegration complex through the intact nuclear membrane (Springett et al. 1989; Miller et al. 1990; Roe et al.

1993). Therefore, culture conditions are needed that support hematopoietic stem cell divisions without differentiation. Also, retroviral vectors depend on expression of the appropriate receptor on the target cell surface. In the following, these two prerequisites of retroviral gene transfer into hematopoietic stem cells will be discussed.

10.3 Ex Vivo Cell Cycling of Hematopoietic Stem Cells as a Prerequisite for Retroviral Gene Transfer

Ex vivo culture of stem cells enables manipulation of the graft in the context of autologous hematopoietic stem cell transplantation. Apart from gene therapy, techniques to refine transplants through ex vivo culture include removal of tumor cells (purging), multiplication of the number of stem cells (stem cell expansion), and improvement of transplant performance in the hematopoietic reconstitution of one or all cell lineages (prestimulation, expansion of committed progenitor cells). The identification of conditions that efficiently maintain the survival or even induce the proliferation of stem cells in culture is one of the key challenges in the field of modern hematopoiesis.

Many researchers have used mice as an in vivo model to investigate the mechanisms involved in hematopoietic stem cell ex vivo expansion. Two decades ago, ex vivo maintenance of hematopoietic stem cells was shown by reconstitution of the bone marrow of lethally irradiated recipient animals with murine long-term bone marrow cultures (Dexter et al. 1977; Spooncer and Dexter 1983). By retrovirus-mediated DNA marking it was demonstrated that, during the long-term bone marrow culture period, repopulating stem cells divide, while the overall number of competitive repopulating units (CRUs) of stem cells decline over time (Fraser et al. 1992). Muench et al. (1993) demonstrated the that ex vivo stimulation by cytokines in culture could positively affect the speed of bone marrow regeneration. The recovery of peripheral blood cell counts was accelerated in lethally irradiated mice receiving transplants of expanded bone marrow [7 days in the presence of interleukin-1 (IL-1) and c-kit ligand] relative to control mice receiving transplants of fresh bone marrow. Similar results were obtained in a study by Han et al. (1993). More recently, Brown et al. (1997) showed an expansion of murine colony-forming units-granulocyte macrophage (CFU-GM) using a serum-free culture system supplemented with SCF and GM-CSF. More primitive spleen-colony forming units (CFU-S) were only maintained, and the repopulation potential of the expanded cells was lost after 1 week of culture. In different studies, using highly enriched stem cell populations instead of whole bone marrow, only phenotypical, but not functional, expansion of the highly enriched stem cell fractions was achieved (Rebel et al. 1994; Spangrude et al. 1995). Taken together, these studies suggest that, in the murine system, an effect on the acceleration of hematopoietic reconstitution may be obtained by cytokine-mediated ex vivo expansion of bone marrow cells. An expansion of long-term repopulating cells was not achieved, and the duration of their maintenance was limited with the methods used.

Initially, the most commonly used method for ex vivo culture of human hematopoietic cells was the "Dexter-type" long-term bone marrow culture (Dexter et al. 1977; Gartner and Kaplan 1980), in which a pre-established stromal layer supports the survival, proliferation and differentiation of progenitor cells over many weeks and maintains a fraction the of bone marrow-reconstituting stem cells. Eaves and co-workers introduced long-term stromal culture for clinical use in the treatment of chronic myelogenous leukemia (CML), showing a partial maintenance of repopulating stem cells in humans (Eaves et al. 1993). In many studies, stromal support has been indispensable for the maintenance of hematopoietic stem cells (Dexter et al. 1984; Verfaillie 1992; Gan et al. 1997; Koller et al. 1997). More recently, Dao et al. (1997) have provided evidence that recombinant growth factors can substitute in part for stromal function in suspension culture. After transplantation of ex vivo cultured and retrovirally marked human hematopoietic cells into immunodeficient mice, vector provirus was detectable in the marrow recovered from nine of ten mice transplanted with human CD34+ cells transduced with stromal support, five of 11 mice that had received human cells transduced in suspension culture with flt-3 ligand, but none of the ten mice that had received human cells transduced in suspension culture without flt-3 ligand. This finding is of special interest because the need for stroma poses considerable problems for clinical applications. The use of autologous stroma requires an additional invasive harvesting procedure to obtain patient bone marrow. In addition, the autologous stroma introduces a highly variable factor into the system. Primary stroma of allogeneic donors has, as does any donor-derived

Table 1. Expansion of primitive human hematopoietic progenitor cells

Study	Starting cells	Cytokines supplemented (dosage)	Culture conditions and duration	In vitro results	In vivo results
Koller et al. 1993	Bone marrow MNC	IL-3 (2 ng/ml); GM-CSF (5 ng/ml); SCF (10 ng/ml); Epo (0.1 U/ml)	10% HS/10% FBS Continuous perfusion cultures 14 days	LTC-IC Expansion 7.5-fold	–
Petzer et al. 1996a	Bone marrow CD34+/CD38- cells	IL-3 ng/ml; IL-6 ng/ml; SCF (ng/ml); flt-3 (ng/ml) G-CSF (ng/ml)	Serum-free/single cell cultures 10 days	Expansion 30-fold	–
Zandstra et al. 1997	LTC-IC Bone marrow CD34+/CD38- cells	IL-3 (60 ng/ml); SCF (300 ng/ml); flt-3 (300 ng/ml)	Serum-free/single cell cultures 10 days	LTC-IC Expansion 45-fold	–
Bhatia et al. 1997	Bone marrow	MIP-1- (ng/ml); IL- 3 (ng/ml); SCM	Serum-containing/ stroma-noncontact cultures 8 weeks	LTC-IC Expansion 5-fold	–
Barnett et al. 1994	Bone marrow	None	Autologous stroma Support	LTC-IC, 570 cells/kg after culture maintenance	Sustained, ph^+ hematopoietic recovery in humans
Henschler et al. 1994	Peripheral blood	IL- 1 (ng/ml); IL-3 (ng/ml); IL- 6 (ng/ml); SCF (ng/ml); Epo (U/ml)	Autologous plasma (2%)/suspension culture 2 weeks	LTC-IC Maintenance	–
Petzer et al. 1997	Peripheral blood	IL-3 (20 ng/ml); G-CSF (20 ng/ml): SCF (100 ng/ml); flt-3 (100 ng/ml)	Serum-free/ suspension culture 5–8 days	LTC-IC Expansion 2–5-fold	–
Brugger et al. 1995	Peripheral blood	IL- 1 (3 ng/ml); IL-3 (100 ng/ml); IL- 6 (100 ng/ml); SCF (10 ng/ml);	Autologous plasma (2%)/suspension culture 12–14 days	–	Rapid hematopoietic recovery in humans
Piacibello et al. 1997	Cord blood $CD34^+$	flt-3 (50 ng/ml); TPO (10 U/ml)	IMDM/10% FCS Suspension culture 2 weeks	LTC-IC Expansion 160-fold	–
Conneally et al. 1996	Cord blood CD34+/CD38-	IL-3 (20 ng/ml); IL-6 (20 ng/ml); G-CSF (20 ng/ml); SCF (100 ng/ml); flt-3 (100 ng/ml)	Serum-free/ suspension culture 5–8 days	LTC-IC Expansion 5-fold	NOD-SCID CRU Expansion 2-fold
Bhatia et al. 1997	Cord blood $CD34^+/CD38^+$	Not specified	Serum-free/ suspension culture 4 days	–	NOD-SCID CRU Expansion 2–4-fold

MNC,mononuclear cells; IL, interleukin; GM-CSF, granulocyte/macrophage colony-stimulating factor; SCF, c-kit ligand (stem cell factor); Epo, erythropoietin; HS, horse serum; FBS, fetal bovine serum; LTC-IC, very immature human hematopoietic progenitor cells; flt, flt-3 ligand; MIP, macrophage inflammatory protein; SCM, stroma-conditioned medium; TPO, thrombopoietin; IMDM, Iscove's modified Dulbecco's medium; NOD, nonobese diabetic mouse; SCID, severe combined immunodeficiency disease mouse; CRU, competitive repopulating units.

tissue, disadvantages with respect to biosafety. The extent to which primary stromal cells can be expanded is limited. In addition, due tothe semi-adherent nature of immature progenitor and stem cells on adherent stromal cell layers, their harvest at the end of the culture period is technically difficult. It requires mechanical and/or enzymatic mobilization, leading to the reinfusion of stromal cells and detritus into the patient. Immortalized human stromal cell lines have been described which support human hematopoiesis in vitro and are potentially unlimited in quantity (Roecklein and Torok-Storb 1995). However, no constant source of allogeneic stroma support is currently available for clinical use. More recently, an automatic device has been introduced that allows the simultaneous growth of stromal and hematopoietic cells of human autologous marrow from a biopsy size sample into a complete marrow transplant in a continuous perfusion system (Stiff et al. 1997).

For these reasons, the ex vivo culture of hematopoietic stem cells in suspension without stroma has been actively pursued by many investigators. Table 1 summarizes the data from reports focusing on ex vivo expansion of very immature human hematopoietic progenitor cells (LTC-IC), of human in vivo repopulating stem cells in xenotransplant animal models or of human hematopoietic repopulation in clinical protocols. Several conclusions can be drawn from the available human stem cell expansion studies. (1) Cytokines which influence cell cycling and differentiation of very immature hematopoietic progenitor cells are flt-3 ligand, stem cell factor (c-kit ligand, SCF), thrombopoietin (TPO, PDGF), erythropoietin (Epo), IL-3 and IL-6. (2) A lack of serum in the ex vivo culture leads to a higher expansion of very primitive hematopoietic progenitor and stem cells. (3) Evidence from xenotransplant models and the first clinical studies (Brugger et al. 1995) suggests that ex vivo expansion of human hematopoietic stem cells may be possible. (4) Gene marking studies are required to prove ex vivo self-renewal of human in vivo repopulating stem cells (Nolta et al. 1996).

10.4 Retroviral Receptor Expression

Retroviral vectors adsorb to the target cells through specific receptors. Human Pit 2 acts as the receptor for the amphotropic MLV envelope. Therefore, the level of Pit 2 expression may be an additional limiting factor for gene transfer into hematopoietic stem cells with MLV-derived retroviral vectors. Orlic et al. (1996) found only low levels of human Pit 2 mRNA in purified populations of primitive human CD34+/CD38— hematopoietic progenitor cells and higher levels in committed CD34+/CD38+ progenitor cells. They postulate that the low level of gene transfer efficiency of amphotropic retroviral vectors into human hematopoietic stem cells reflects the low level of Pit 2 expression in the relevant primitive hematopoietic population. Several approaches are currently under investigation to overcome the problem of low receptor expression, including attempts to induce an up-regulation of the Pit family receptors or to overexpress them by transient expression vectors in the target cells. MLV-based vectors have been pseudotyped with alternative envelope proteins to target different viral receptors.

The receptors of the Pit family function as sodium-dependent phosphate symporters (Kavanaugh et al. 1994). Therefore, depletion of phosphate from the extracellular medium may result in an up-regulation of the receptor molecules. Bunnell et al. (1995) showed a threefold increase in transduction efficiency of primary human lymphocytes by phosphate depletion. Whether phosphate depletion is applicable to the transduction of hematopoietic stem cells is not yet known. In another approach to increase the expression level of retroviral receptors, recombinant adenovirus receptors were constructed which encode for Pit 2. In Hela cells, Lieber et al. (1995) demonstrated a tenfold increase in gene transfer efficiency by amphotropic retroviral vectors after transient expression of the appropriate receptor through adenovirus-mediated transduction. It seems likely that the same principle may improve gene transfer into hematopoietic stem cells because it has been shown that hematopoietic cells can be transduced by adenoviral vectors (Mitani et al. 1994; Neering et al. 1996).

Currently, vectors from one of the most promising pseudotyped MLV-based packaging systems enters the target cells via the gibbon ape leukemia virus (GALV) receptor Pit 1 instead of the amphotropic murine leukemia virus (MLV) receptor Pit 2 (Miller et al. 1991). It has been shown that in bone marrow Pit 1 is expressed at higher level than Pit 2 (Kavanaugh et al. 1994). Consistently, the GALV pseudotyped vector PG13/LN allows for a higher transduction efficiency of committed human hematopoietic progenitor cells (von Kalle et al. 1994). Recent studies directly comparing gene transfer efficiency into baboon repopulating stem cells provide strong evidence that PG13 may be superior to amphotropic vectors for the purpose of primate stem cell transduction (H.P. Kiem, personal communication).

10.5 Preclinical Studies

The proof of principle that retroviral transduction of self-renewing hematopoietic stem cells is possible in mammals was worked out in the murine system more than a decade ago (Dick et al. 1985; Bodine et al. 1989; Hollander et al. 1992). Dick et al. (1985) were the first to demonstrate that gene transfer into all repopulating stem cells of an animal could be achieved. They combined mobilization of hematopoietic stem cells by myelotoxic therapy, cytokine stimulation during transduction culture and in vitro selection of transduced stem cells. Later, it was shown that the progeny of a single murine hematopoietic stem cell can sufficiently repopulate all three hematopoietic lineages of a mouse (Lemischka et al. 1986). In contrast to the encouraging mouse data, the results of gene transfer studies in large outbred animals were less convincing. Retroviral transduction of repopulating hematopoietic stem cells has been shown in dogs, cats, sheep and monkeys, but at a transduction efficiency of only 5% or less (Stead et al. 1988; Kantoff et al. 1989; Lothrop et al. 1991; Carter et al. 1992; van Beusechem et al. 1992; Bodine et al. 1993). In dogs, the utility of different hematopoietic stem cell sources has been compared showing equal efficiency of transduced peripheral blood stem cell (PBSC) transplants and transduced marrow cells for obtaining hematopoietic recovery after otherwise lethal total body irradiation (Kiem et al. 1994). By genetic marking in 1%–2% of marrow CFU-GM long after transplantation, it was proven that cytokine-mobilized PBSCs contained long-term repopulating hematopoietic stem cells.

10.6 Clinical Studies

With one exception, all of the clinical studies of gene transfer into hematopoietic stem cells that have been concluded thus far were marking studies using MLV derived retroviral vectors encoding the neomycin phosphotransferase gene as a marker gene. The only study reported to date of a therapeutic gene transfer is a cord blood study in ADA-deficient children (Kohn et al. 1995). The other studies were Phase I studies designed to test the feasibility and toxicity of detecting contaminating tumor cells in autologous marrow or PBSC transplants in acute myeloid leukemia (AML), neuroblastoma, CML and breast cancer by genetic marking (Brenner et al. 1993a,b; Deisseroth et al. 1994; Bordignon et al. 1995; Dunbar et al. 1995). Dunbar et al. (1995) compared retrovirally marked bone marrow derived stem cells (BMSCs) to PBSCs with respect to their ability to reconstitute hematopoiesis in patients after high-dose chemotherapy. After cotransplantation of 11 multiple myeloma or breast cancer patients with marked hematopoietic cells of both sources, in all ten patients that could be evaluated marked cells could be demonstrated with a tendency of better long-term marking of the hematopoiesis originating from the bone marrow graft. For most of the malignant diseases tested, gene marking proved that contaminating tumor cells from the autologous stem cell transplants contributed to disease relapses. Moreover, the experience from the studies is proof of principle that retroviral gene transfer and long-term persistence of the transgene in human hematopoietic stem cells can be achieved without detectable side effects. Gene transfer efficiency was rather disappointing, ranging from less than 1/100 to 1/100 000, indicating that only a small number of repopulating CD34 BMSCs and PBSCs underwent cell division and retroviral transduction during the ex vivo culture period. There may be several explanations for this observation: The short-term suspension cultures used may not support ex vivo cycling of hematopoietic stem cells without subsequent cellular differentiation, i.e., without loss of long-term repopulating capability. More efficient gene transfer protocols will either have to obtain efficient self renewing cell divisions of hematopoietic stem cells, for example by elongating the transduction period (von Kalle et al. 1994) or introducing new cytokines with a proliferative activity in stem cells like the flt-3 ligand (Petzer et al. 1996b; Zandstra et al. 1997), or will have to circumvent the requirement for stem cell cycling by using a vector capable of gene transfer into nondividing cells. Moreover, follow-up reports from patients that had received retrovirally marked T cells in the first pioneering clinical trials indicate that immunocompetent individuals mount a cellular immune response against the genetically modified cells, probably induced by the

MHC class I related expression of the genetically transferred neo-antigen. Currently, retroviral vectors are being developed in several groups that incorporate the cytomegtalovirus (CMV) US II gene or other viral sequences intended to suppress MHC class I expression and thus T cell activation by the genetically engineered cells.

10.7 Future Perspectives

To circumvent the need for ex vivo cell cycling of hematopoietic stem cells to obtain efficient gene therapy of the hematopoietic system, much effort has been made to develop alternative vectors derived from viruses that infect and stably integrate in nondividing cells. New vectors based on AAV or human foamy virus (HFV) may have clear advantages over MLV vectors with regard to tropism and titer, whereas their capability of infecting nondividing cells is still a matter of debate (Podsakoff et al. 1994; Bieniasz et al. 1995; Halbert et al. 1995; Fisheradams et al. 1996; Hirata et al. 1996; Russell and Miller 1996). The most promising new viral vector developments are based on lentiviruses such as simian (SIV) and human immunodeficiency virus (HIV), because there is strong evidence for the capability of wild-type lentiviruses to infect nondividing cells. The integrated proviral cDNA of these lentiviruses can be found in the nucleus of terminally differentiated, nondividing cells in a newly infected host. Naldini et al. (1996) developed first generation vector systems from lentiviruses that allowed efficient gene transfer in nondividing neurons. Now, helper virus-free packaging cell lines will have to be developed. In vectors derived from viruses with known pathogenicity in humans, e.g., AAV and HIV, wild-type virus infection of a vector-treated individual could lead to mobilization of vector sequences by recombinations of wild-type virus with vector sequences. Thus considerable further development is necessary to achieve a standard of biological safety sufficient for clinical application.

10.8 Conclusions

Gene transfer into hematopoietic stem cells is one of the most intriguing yet elusive goals of gene therapy. Recent advances in the ex vivo manipulation of hematopoietic stem cells as well as in the design of integrating vector systems and their packaging have moved hematopoietic gene transfer at therapeutically relevant levels closer to becoming a reality.

Acknowledgements. Supported by grants Ka 976/1-1, 976/4-1 awarded by the Deutsche Forschungsgemeinschaft and 01KV9527 awarded by the German Ministery of Education and Research.

References

Barnett MJ, Eaves CJ, Phillips GL, Gascoyne RD, Hogge DE, Horsman DE, Humphries RK, Klingemann HG, Lansdorp PM, Nantel SH et al (1994) Autografting with cultured marrow in chronic myeloid leukemia: results of a pilot study (see comments). Blood 84:724–32

Bhatia R, Mcglave PB, Miller JS, Wissink S, Lin WN, Verfaillie CM (1997) A clinically suitable ex vivo expansion culture system for ltc-ic and cfc using stroma-conditioned medium. Exp Hematol 25:980–991

Bieniasz PD, Weiss RA, Mcclure MO (1995) Cell cycle dependence of foamy retrovirus infection. J Virol 69:7295–7299

Bodine DM, Karlsson S, Papayannopoulou T, Nienhuis AW (1989) Expression of human beta globin genes introduced into primitive murine hematopoietic progenitor cells by retrovirus mediated gene transfer. Prog Clin Biol Res 316B:219–233

Bodine DM, Moritz T, Donahue RE, Luskey BD, Kessler SW, Martin DI, Orkin SH, Nienhuis AW, Williams DA (1993) Long-term in vivo expression of a murine adenosine deaminase gene in rhesus monkey hematopoietic cells of multiple lineages after retroviral mediated gene transfer into CD34+ bone marrow cells. Blood 82:1975–1980

Bordignon C, Notarangelo LD, Nobili N, Ferrari G, Casorati G, Panina P, Mazzolari E, Maggioni D, Rossi C, Servida P, Ugazio AG, Mavilio F (1995) Gene therapy in peripheral blood lymphocytes and bone marrow for ADA(-) immunodeficient patients. Science 270:470–475

Brenner MK, Rill DR, Holladay MS, Heslop HE, Moen RC, Buschle M, Krance RA, Santana VM, Anderson WF, Ihle JN (1993a) Gene marking to determine whether autologous marrow infusion restores long-term haemopoiesis in cancer patients. Lancet 342:1134–1137

Brenner MK, Rill DR, Moen RC, Krance RA, Mirro JJ, Anderson WF, Ihle JN (1993b) Gene-marking to trace origin of relapse after autologous bone-marrow transplantation. Lancet 341:85–86

Brown RL, Sheng F, Dusing SK, Fischer R, Patchen M (1997) Serum-free culture conditions for cells capable of producing long-term survival in lethally irradiated mice. Stem Cells 15:237–245

Brugger W, Heimfeld S, Berenson RJ, Mertelsmann R, Kanz L (1995) Reconstitution of hematopoiesis after high-dose chemotherapy by autologous progenitor cells generated ex vivo. N Engl J Med 333:283–287

Bunnell BA, Muul LM, Donahue RE, Blaese RM, Morgan RA (1995) High-efficiency retroviral-mediated gene transfer into human and nonhuman primate peripheral blood lymphocytes. Proc Natl Acad Sci USA 92:7739–7743

Carter RF, Abrams OA, Dick JE, Kruth SA, Valli VE, Kamel RS, Dube ID (1992) Autologous transplantation of canine long-term marrow culture cells genetically marked by retroviral vectors. Blood 79:356–364

Conneally E, Cashman J, Petzer AL, Eaves CJ (1996) In vitro expansion of human lympho-myeloid stem cells from cord blood demonstrated using a quantitative in vivo repopulating assay. Blood

Dao MA, Hannum CH, Kohn DB, Nolta JA (1997) Flt3 ligand preserves the ability of human cd34(+) progenitors to sustain long-term hematopoiesis in immune-deficient mice after ex vivo retroviral-mediated transduction. Blood 89:446–456

Deisseroth AB, Zu Z, Claxton D, Hanania EG, Fu S, Ellerson D, Goldberg L, Thomas M, Janicek K, Anderson WF, Hester J, Korbling M, Durett A, Moen R, Berenson R, Heimfeld S, Hamer J, Clavert L, Tibbits P, Talpaz M, Kantarijan H, Champlin R, Reeding C (1994) Genetic marking shows that Ph+ cells present in autologous transplants of chronic myelogenous leukemia (CML) contribute to relapse after autologous bone marrow in CML. Blood 83:3068–3076

Dexter TM, Allen TD, Lajtha LG (1977) Conditions controlling the proliferation of hematopoietic cells in vitro. J Cell Physiol 91:335–344

Dexter TM, Simmons P, Purnell RA, Spooncer E, Schofield R (1984) The regulation of hemopoietic cell development by the stromal cell environment and diffusible regulatory molecules. Prog Clin Biol Res 148:13–33

Dick JE, Magli MC, Huszar D, Phillips RA, Bernstein A (1985) Introduction of a selectable gene into primitive stem cells capable of long-term reconstitution of the hemopoietic system of W/Wv mice. Cell 42:71–79

Dunbar CE, Cottlerfox M, Oshaughnessy JA, Doren S, Carter C, Berenson R, Brown S, Moen RC, Greenblatt J, Stewart FM, Leitman SF, Wilson WH, Cowan K, Young NS, Nienhuis AW (1995) Retrovirally marked CD34-enriched peripheral blood and bone marrow cells contribute to long-term engraftment after autologous transplantation. Blood 85:3048–3057

Eaves AC, Eaves CJ, Phillips GL, Barnett MJ (1993) Culture purging in leukemia: past, present, and future. Leuk Lymph 11:259–263

Fisheradams G, Wong KK, Podsakoff G, Forman SJ, Chatterjee S (1996) Integration of adeno-associated virus vectors in cd34(+) human hematopoietic progenitor cells after transduction. Blood 88:492–504

Fraser CC, Szilvassy SJS, Eaves CJ, Humphries RK (1992) Proliferation of totipotent hematopoietic stem cells in vitro with retention of long-term competitive in vivo reconstituting ability. Proc Natl Acad Sci USA 89:1968–1972

Gan OI, Murdoch B, Larochelle A, Dick JE (1997) Differential maintenance of primitive human scid-repopulating cells, clonogenic progenitors, and long-term culture-initiating cells after incubation on human bone marrow stromal cells. Blood 90:641–650

Gartner S, Kaplan HS (1980) Long-term culture of human bone marrow cells. Proc Natl Acad Sci USA 77:4756–4759

Halbert CL, Alexander IE, Wolgamot GM, Miller AD (1995) Adeno-associated virus vectors transduce primary cells much less efficiently than immortalized cells. J Virol 69:1473–1479

Han M, Kobayashi M, Imamura M, Hashino S, Kobayashi H, Maeda S, Iwasaki H, Fujii Y, Musashi M, Sakurada K et al (1993) In vitro expansion of murine hematopoietic progenitor cells in liquid cultures for bone marrow transplantation: effects of stem cell factor. Int J Hematol 57:113–120

Henschler R, Brugger W, Luft T, Frey T, Mertelsmann R, Kanz L (1994) Maintenance of transplantation potential in ex vivo expanded CD34(+)-selected human peripheral blood progenitor cells. Blood 84:2898–2903

Hirata RK, Miller AD, Andrews RG, Russell DW (1996) Transduction of hematopoietic cells by foamy virus vectors. Blood 88:3654–3661

Hollander GA, Luskey BD, Williams DA, Burakoff SJ (1992) Functional expression of human CD8 in fully reconstituted mice after retroviral-mediated gene transfer of hemopoietic stem cells. J Immunol 149:438–444

Kantoff PW, Flake AW, Eglitis MA, Scharf S, Bond S, Gilboa E, Erlich H, Harrison MR, Zanjani ED, Anderson WF (1989) In utero gene transfer and expression: a sheep transplantation model. Blood 73:1066–1073

Kavanaugh MP, Miller DG, Zhang W, Law W, Kozak SL, Kabat D, Miller AD (1994) Cell-surface receptors for gibbon ape leukemia virus and amphotropic murine retrovirus are inducible sodium-dependent phosphate symporters. Proc Natl Acad Sci USA 91:7071–7075

Kiem HP, Darovsky B, von Kalle C, Goehle S, Stewart D, Graham T, Hackman R, Appelbaum FR, Deeg HJ, Miller AD et al (1994) Retrovirus-mediated gene transduction into canine peripheral blood repopulating cells. Blood 83:1467–1473

Kohn DB, Weinberg KI, Nolta JA, Heiss LN, Lenarsky C, Crooks GM, Hanley ME, Annett G, Brooks JS, Elkhoureiy A, Lawrence K, Wells S, Moen RC, Bastian J, Williamsherman DE, Elder M, Wara D, Bowen T, Hershfield MS, Mullen CA, Blaese RM, Parkman R (1995) Engraftment of gene-modified umbilical cord blood cells in neonates with adenosine deaminase deficiency. Nat Med 1:1017–1023

Koller MR, Emerson SG, Palsson BO (1993) Large-scale expansion of human stem and progenitor cells from bone marrow mononuclear cells in continuous perfusion cultures. Blood 82:378–384

Koller MR, Manchel I, Palsson BO (1997) Importance of parenchymal/stromal cell ratio for the ex vivo reconstitution of human hematopoiesis. Stem Cells 15:305–313

Lemischka IR, Raulet DH, Mulligan RC (1986) Developmental potential and dynamic behavior of hematopoietic stem cells. Cell 45:917–927

Lieber A, Vrancken-Peeters MTFD, Kay MA (1995) Adenovirus-mediated transfer of the amphotropic retrovirus receptor cDNA increases retroviral transduction in cultured cells. Hum Gene Ther 6:5–11

Lothrop CJ, Niemeyer GP, Jones JB, Peterson MG, Smith JR, Baker HJ, Morgan RA, Eglitis MA, Anderson WF (1991) Expression of a foreign gene in cats reconstituted with retroviral vector infected autologous bone marrow. Blood 78:237–245

Miller AD, Rosman GJ (1989) Improved retroviral vectors for gene transfer and expression. Biotechniques 7:980–982

Miller DG, Adam MA, Miller AD (1990) Gene transfer by retrovirus vectors occurs only in cells that are actively replicating at the time of infection. Mol Cell Biol 10:4239–4242

Miller AD, Garcia JV, von Suhr N, Lynch CM, Wilson C, Eiden MV (1991) Construction and properties of retrovirus packaging cells based on gibbon ape leukemia virus. J Virol 65:2220–2224

Mitani K, Graham FL, Caskey CT (1994) Transduction of human bone marrow by adenoviral vector. Hum Gene Ther 5:941–948

Muench MO, Firpo MT, Moore MAS (1993) Bone marrow transplanation with interleukin-1 plus kit-ligand ex vivo expanded bone marrow accelerates hematopoietic reconstitution in mice without loss of stem cell lineage and proliferative potential. Blood 81:3463–3473

Naldini L, Blomer U, Gallay P, Ory D, Mulligan R, Gage FH, Verma IM, Trono D (1996). In vivo gene delivery and stable transduction of nondividing cells by a lentiviral vector. Science 272:263–267

Neering SJ, Hardy SF, Minamoto D, Spratt SK, Jordan CT (1996) Transduction of primitive human hematopoietic cells with recombinant adenovirus vectors. Blood 88:1147–1155

Nienhuis AW (1994) Gene transfer into hematopoietic stem cells. Blood Cells 20:141–150

Nolta JA, Dao MA, Wells S, Smogorzewska EM, Kohn DB (1996) Transduction of pluripotent human hematopoietic stem cells demonstrated by clonal analysis after engraftment in immune-deficient mice. Proc Natl Acad Sci USA 93:2414–2419

Orlic D, Girard LJ, Jordan CT, Anderson SM, Cline AP, Bodine DM (1996) The level of mrna encoding the amphotropic retrovirus receptor in mouse and human hematopoietic stem cells is low and correlates with the efficiency of retrovirus transduction. Proc Natl Acad Sci USA 93:11097–11102

Petzer AL, Eaves CJ, Barnett MJ, Eaves AC (1997) Selective expansion of primitive normal hematopoietic cells in cytokine-supplemented cultures of purified cells from patients with chronic myeloid leukemia. Blood 90:64–69

Petzer AL, Hogge DE, Lansdorp PM, Reid DS, Eaves CJ (1996a) Self-renewal of primitive human hematopoietic cells (long-term-culture-initiating cells) in vitro and their expansion in defined medium. Proc Natl Acad Sci USA 93:1470–1474

Petzer AL, Zandstra PW, Piret JM, Eaves CJ (1996b) Differential cytokine effects on primitive (cd34+cd38(-)) human hematopoietic cells – novel responses to flt3-ligand and thrombopoietin. J Exp Med 183:2551–2558

Piacibello W, Sanavio F, Garetto L, Severino A, Bergandi D, Ferrario J, Fagioli F, Berger M, Aglietta M (1997) Extensive amplification and self-renewal of human primitive hematopoietic stem cells from cord blood. Blood 89:2644–2653

Podsakoff G, Wong KJ, Chatterjee S (1994) Efficient gene transfer into nondividing cells by adeno-associated virus-based vectors. J Virol 68:5656–5666

Rebel VI, Dragowska W, Eaves CJ, Humphries RK, Lansdorp PM(1994) Amplification of Sca-1+ Lin- WGA+ cells in serum-free cultures containing steel factor, interleukin-6, and erythropoietin with maintenance of cells with long-term in vivo reconstituting potential. Blood 83:128–136

Roe T, Reynolds T, Yu G, Brown P (1993) Integration of murine leukemia virus DNA depends on mitosis. EMBO 12:2099

Roecklein BA, Torok-Storb B (1995) Functionally distinct human marrow stromal cell lines immortalized by transduction with the human papilloma virus e6/e7 genes. Blood 85:997–1005

Russell DW, Miller AD (1996) Foamy virus vectors. J Virol 70:217–222

Spangrude GJ, Brooks DM, Tumas DB (1995) Long-term repopulation of irradiated mice with limiting numbers of purified hematopoietic stem cells: in vivo expansion of stem cell phenotype but not function. Blood 85:1006–1016

Spooncer E, Dexter TM (1983) Transplantation of long-term cultured bone marrow cells. Transplantation 35:624–627

Springett GM, Moen RC, Anderson S, Blaese RM, Anderson WF (1989) Infection efficiency of T lymphocytes with amphotropic retroviral vectors is cell cycle dependent. J Virol 63:3865–3869

Stead RB, Kwok WW, Storb R, Miller AD (1988) Canine model for gene therapy: inefficient gene expression in dogs reconstituted with autologous marrow infected with retroviral vectors. Blood 71:742–747

Stiff PJ, Oldenberg d, Hsi E, Chen B, Douville J, Burhop S, Bayer R, Peace D, Malhotra D, Kerger C, Armstrong D, Muller T (1997) Successful hematopoietic engraftment following high dose chemotherapy using only exvivo expandedbone marrow grown in stromal based Bioreactors. ASCO Meeting, 17–20 May 1997, Denver CO (abstract)

van Beusechem VW, Kukler A, Heidt PJ, Valerio D (1992) Long-term expression of human adenosine deaminase in rhesus monkeys transplanted with retrovirus-infected bone-marrow cells. Proc Natl Acad Sci USA 89:7640–7644

Verfaillie CM (1992) Direct contact between human primitive hematopoietic progenitors and bone marrow stroma is not required for long-term in vitro hematopoiesis. Blood 79:2821–2826

von Kalle C, Kiem HP, Goehle S, Darovsky B, Heimfeld S, Torok SB, Storb R, Schuening FG (1994) Increased gene transfer into human hematopoietic progenitor cells by extended in vitro exposure to a pseudotyped retroviral vector. Blood 84:2890–2897

Zandstra PW, Conneally E, Petzer AL, Piret JM, Eaves CJ (1997) Cytokine manipulation of primitive human hematopoietic cell self-renewal. Proc Natl Acad Sci USA 94:4698–4703

11 Immunological Approaches for Gene Therapy of Cancer

R.E. Sobol, D. Shawler, C. Van Beveren, M. Garrett, H. Fakhrai, R. Bartholomew, I. Royston, and D.P. Gold

11.1 Introduction

The identification of immunostimulatory and tumor antigen genes combined with advances in our ability to modify gene expression has fostered a new era of tumor immunotherapy. These novel immuno-gene therapies include tumor cell vaccines genetically engineered to express cytokine genes or modified by antisense vectors to inhibit immunosuppressive or differentiation factors, cytokine gene transfer into tumor

Table 1. Immuno-gene therapy clinical trials worldwide

Gene	Number of protocols
Cytokines	
Interleukin-2	28
Granulocyte/macrophage-colony stimulating factor	8
Interleukin-12	4
Interferon-γ	2
Tumor necrosis factor-α	2
Interleukin-7	2
Interleukin-2/interferon-γ	1
Tumor antigens	
Carcinoembryonic antigen	6
Prostate-specific antigen	3
MART (melanoma antigen recognized by T cells)	3
Ig idiotype	2
gp100	2
Other	
HLA-B7/B_2micro	13
CD80 (B7.1)	5
T cell receptor	3
Carcinoembryonic antigen/CD80	2
Insulin-like growth factor-1 antisense	1
Transforming growth factor- antisense	1

infiltrating lymphocytes (TILs), intratumoral injection of allogeneic MHC or cytokine cDNA and vaccination with tumor antigen nucleic acids. Immuno-gene therapy is the most frequent form of gene therapy in current clinical trials. Table 1 lists the clinical trials which have been submitted for approval to regulatory agencies worldwide. Each of these novel immuno-gene therapies has advantages and disadvantages with respect to their potential for clinical applications. In general, immunotherapies involving autologous tumor cells have the potential advantage of containing the largest number of pertinent tumor antigens for a particular patient. However, the customized nature of autologous tumor and TIL-based treatments make them expensive and difficult to manufacture. The use of allogeneic cell lines in vaccine preparations provides practical advantages. However, the efficacy of allogeneic tumor vaccines requires the shared expression of tumor antigens by the vaccine and the patient's tumor. Similarly, the efficacy of nucleotide vaccines encoding tumor antigens requires diffuse tumor expression of a target antigen capable of mediating tumor cell lysis. Intratumoral injection of immunostimulatory genes circumvents laborious ex vivo cultures or antigen matching but is problematic in the adjuvant setting to treat microscopic metastases following surgical resection of the primary tumor. It is generally recognized that immunotherapies are most effective when the tumor burden is small and immuno-gene therapies may be most efficacious in the adjuvant setting following surgical or radiation treatments.

The results of the clinical trials listed in Table 1 should identify the modalities, target antigens, cytokines and immunomodulatory gene modifications which should be further developed for future clinical applications. Our laboratory has been pursuing the development of tumor cell therapeutic vaccines for colorectal cancer. Our pre-clinical and clinical efforts in this area are summarized in the remainder of this review.

11.2 Immunotherapy of Colorectal Carcinoma

Colorectal carcinoma is one of the most common cancers in the United States with an annual incidence of greater than 150 000 individuals. Most patients are treated with tumor resection and do not have clinically detectable tumor following surgery. However, a significant number of these patients have microscopic metastases and eventually relapse with clinically overt disease in the liver or abdominal cavity. The large number of patients and the small tumor burden following surgical resection makes colorectal carcinoma an attractive candidate for adjuvant immunotherapy trials. It is generally acknowledged that immunotherapies are likely to be most effective when the tumor burden is low. In this regard, the immunomodulator levamisole is currently approved for the treatment of patients with Duke's C tumors (metastases to abdominal lymph nodes). In addition, encouraging results have been obtained with an autologous tumor vaccine as an adjuvant therapy following tumor resection (Hoover et al. 1993). In this study, immunization with autologous tumor preparations and the adjuvant BCG resulted in a significant increase in disease free and total survival (Hoover et al. 1993). Additional studies have suggested therapeutic efficacy of passive and active immunotherapies directed against the tumor associated antigens (TAAs) recognized by the monoclonal antibodies 17-1A and GA733 (Herlyn et al. 1991, 1994). These findings indicate that immunotherapies may have beneficial effects in colon carcinoma and support the development of novel immunotherapy approaches for this malignancy.

11.3 Cytokine Immunotherapy

Recent advances in our understanding of the biology of the immune system have led to the identification of numerous cytokines which modulate immune responses (Kelso 1989; Bordon and Sondel 1990). These agents mediate many of the immune responses involved in anti-tumor immunity. Several of these cytokines have been produced by recombinant DNA methodology and evaluated for their anti-tumor effects. In experimental clinical trials, the administration of cytokines and related immunomodulators has resulted in objective tumor responses in some patients with various types of neoplasms (Bordon and Sondel 1990; Rosenberg et al. 1988; Lotze et al. 1986).

Interleukin-2 (IL-2) is an important cytokine in the generation of anti-tumor immunity (Rosenberg et al. 1988). In response to tumor antigens, the helper T cell subset of lymphocytes secretes small quantities of IL-2. This IL-2 acts locally at the site of tumor antigen presentation to activate cytotoxic T cells and natural killer (NK) cells which mediate systemic tumor cell destruction. Intravenous, intralymphatic or intralesional administration of IL-2 has resulted in clinically significant responses in several types of cancer (Rosenberg et al. 1988; Lotze et al. 1986; Pizza et al. 1988; Sarna et al. 1990; Gandolfi et al. 1989). However, severe toxicities (hypotension and edema) limit the dose and efficacy of intravenous and intralymphatic IL-2 administration (Lotze et al. 1986; Sarna et al. 1990). The toxicity of systemically administered cytokines is not surprising since these agents mediate local cellular interactions and they are normally secreted in quantities too small to have systemic effects.

To circumvent the toxicity of systemic IL-2 administration, several investigators have examined intralesional injection of IL-2 (Gandolfi et al. 1989; Bubenik et al. 1988). This approach eliminates the toxicity associated with systemic IL-2 administration. However, multiple intralesional injections are required to optimize therapeutic efficacy (Gandolfi et al. 1989; Bubenik et al. 1988). These injections will be impractical for many patients, particularly when tumor sites are not accessible for direct injection without potential significant morbidity.

Cytokine gene transfer has resulted in significant anti-tumor immune responses in several animal tumor models (Fearon et al. 1990; Gansbacher et al. 1990; Watanabe et al. 1989; Tepper et al. 1989). In these studies, the transfer of cytokine genes into tumor cells has reduced or abrogated the tumorigenicity of the cells after implantation into syngeneic hosts. The transfer of genes for IL-2 (Fearon et al. 1990; Gansbacher et al. 1990), interferon (IFN)-γ (Watanabe et al. 1989), and IL-4 (Tepper et al. 1989) significantly reduced or eliminated the growth of several different histological types of murine tumors. In the studies employing IL-2 gene transfer, the treated animals also developed systemic anti-tumor immunity and were protected against subsequent tumor challenges with the unmodified parental tumor (Fearon et al. 1990; Gansbacher et al. 1990). Similar inhibition of tumor growth and protective immunity were also demonstrated when immunizations were performed with a mixture of unmodified parental tumor cells and genetically modified tumor cells engineered to express the IL-2 gene. No toxicity associated with expression of the cytokine transgenes was reported in these animal tumor studies (Fearon et al. 1990; Gansbacher et al. 1990; Watanabe et al. 1989; Tepper et al. 1989).

11.4 Cytokine Gene Therapy with Genetically Modified Fibroblasts

Unfortunately, many types of tumors are difficult to establish in culture and cytokine gene therapies requiring the transduction of autologous tumor cells may not be practical for many cancer patients. However, primary human fibroblasts obtained from skin biopsies or established allogeneic fibroblast cell lines may be readily cultured in vitro and genetically modified to express and secrete cytokines (Fakhrai et al.

1995; Sobol et al. 1995; Kim and Cohen 1994; Kim et al. 1992; Tahara et al. 1994). The genetically modified irradiated fibroblasts are then mixed with autologous or allogeneic irradiated tumor cells and employed in subcutaneous immunizations to induce systemic anti-tumor immunity. Application of genetically modified fibroblasts in therapeutic vaccines facilitates titration of single or multiple cytokine doses independent of tumor cell doses and permits other forms of genetic manipulation to be performed on the tumor cell component of the vaccines to further enhance their immunogenicity. These considerations provide the rationale for examining the use of autologous or allogeneic fibroblasts genetically modified to secrete cytokines as a means of enhancing anti-tumor immune responses. Several groups, including our own, have demonstrated the efficacy of active tumor immunotherapy with cytokine-transduced syngeneic or allogeneic fibroblasts (Fakhrai et al. 1995; Sobol et al. 1995; Kim and Cohen 1994; Kim et al. 1992; Tahara et al. 1994). In our studies employing a murine colon tumor model, immunizations with a mixture of irradiated tumor cells and irradiated IL-2-transduced fibroblasts generated systemic anti-tumor immunity capable of rejecting a subsequent tumor challenge and eradicating established tumors (Fakhrai et al. 1995). In subsequent work with this same mouse tumor model (Shawler et al. 1997), allogeneic fibroblasts engineered to secrete IL-2 were equally effective at eliciting anti-tumor responses when compared to syngeneic fibroblast secreting IL-2. Similar results of fibroblasts providing the cytokines were observed by Lotze and coworkers in a murine melanoma model following immunizations with tumor cells and allogeneic fibroblasts genetically modified to express IL-12 (Tahara et al. 1994). In related animal studies, Kim and Cohen (1994; Kim et al. 1992) induced systemic anti-tumor immunity following immunization with IL-2 secreting fibroblasts that were also transfected with tumor DNA. In their studies, the results of immunizations with allogeneic and syngeneic fibroblasts transfected with tumor DNA were compared with and without concomitant IL-2 gene transfer. It was found that different types of effector cells were induced by immunizations with IL-2-transduced autologous vs allogeneic fibroblasts and that combined IL-2 gene transfer and allogeneic stimulation had synergistic effects with enhanced survival compared to immunization with either approach alone (Kim and Cohen 1994; Kim et al. 1992). These results support the development of tumor vaccines containing fibroblasts genetically modified to secrete cytokines as a means of enhancing anti-tumor immune responses. The results of pre-clinical animal studies implicate both practical and potential therapeutic advantages to the application of allogeneic fibroblasts for cytokine gene transfer.

11.5 Phase I Clinical Trial of Autologous Fibroblasts Genetically Modified to Secrete Interleukin-2 and Autologous Colon Cancer Cells

We have recently completed a Phase I clinical trial of IL-2 gene therapy in colorectal carcinoma patients comprising injection of irradiated autologous tumor cells mixed with autologous fibroblasts genetically modified to secrete IL-2. In this trial, all patients received 10^7 tumor cells mixed with different doses of IL-2 secreting fibroblasts. Each patient was scheduled to receive three administrations at days 0, 14 and 28. The results of this ten patient trial will be reported in detail elsewhere (Sobol et al., manuscript submitted). Delayed-type hypersensitivity (DTH) skin reactions were observed at the injection sites in five of ten patients. The fact that no patients showed pre-treatment DTH reactions implies that these reactions were induced by the vaccine. Fatigue, flu-like symptoms or fever were noted in eight of ten treated patients although these symptoms were present in four patients prior to the initiation of therapy. There were no significant, treatment related changes in complete blood counts, serum chemistries or urinalyses. Stable disease of 3 months duration was noted in one patient while the remaining patients had progressive disease. Cytotoxic T cell precursor (pCTL) frequency analyses were performed to measure cell-mediated immunity. The patients' autologous tumor cells (ATCs) were utilized as stimulator cells and pre- and post-treatment peripheral blood mononuclear cells (PBMCs) were employed as effector cells in these assays. Low frequencies of tumor pCTLs (range=1/190 000–1/1 320,000 PBMCs) were detected prior to therapy in four of six patients with sufficient cells for evaluation. There was a fivefold increase following treatment in the frequency of tumor pCTLs in two of three evaluable patients with detectable pre-treatment pCTLs. Cloned T cells derived from their corresponding precursor cultures were cytotoxic for tumor cells but not autologous fibroblasts.

11.6 Phase I Clinical Trial of Allogeneic Fibroblasts Genetically Modified to Secrete Interleukin-2 and Autologous Tumor Cells

A Phase I clinical trial has been performed employing the allogeneic KMST-6 fibroblast cell line in a manner similar to our autologous fibroblast study (Veelken et al. 1997). A vaccine composed of autologous tumor cells and IL-2-secreting allogeneic fibroblasts was administered to patients with melanoma and renal cell carcinoma. Autologous tumor cells were isolated from biopsy specimens. A clone (KMST-6.14) of an immortalized human fibroblast line that secreted 5290 IU of IL-2 per 10^6 cells per 24 h was obtained by cationic lipofection with an expression construct for human IL-2 and neomycin resistance [Neo(r)]. Fifteen patients with refractory malignant tumors received three to four injections of irradiated KMST-6.14 and autologous tumor cells. CD8+ T cell lines isolated from vaccination sites of two malignant melanoma patients exhibited lytic activity against autologous tumor cells in vitro. CD8+ T cell clones established from the vaccination site of one of two renal cell carcinoma patients preferentially lysed autologous and partially HLA-matched allogeneic renal cell carcinoma cells. In summary, a vaccine composed of IL-2 gene-transfected allogeneic fibroblasts and autologous tumor cells was able to enhance specific anti-tumor T cell responses in vivo without major side effects (Veelken et al. 1997).

We were encouraged by the results of these Phase I studies in cancer patients treated with vaccines comprised of a mixture of tumors and fibroblasts genetically modified to secrete IL-2. This form of IL-2 gene therapy was well tolerated and appeared to have immunological effects as demonstrated by the observed DTH skin reactions and increased tumor T cell precursor frequencies. These results supported the development of a therapeutic vaccine which employs a more practical approach based upon allogeneic cell lines and includes additional genetic modifications (B7.1/CD80 co-stimulatory gene transfection) to further enhance immunogenicity of the tumor cells. Additional pre-clinical and clinical studies supporting the application of B7.1/CD80 gene modified allogeneic tumor cells for tumor immunotherapy are summarized below.

11.7 Co-stimulatory Molecules and Anti-tumor Immunity

Numerous studies have demonstrated that antigen recognition alone may not be sufficient for optimal activation of T cell-mediated immunity. Second signals such as co-ligation of auxiliary molecules are also critical for generating T cell immune responses (Mondino and Jenkins 1994; June et al. 1990). Antigen recognition in the absence of these second signals can lead to tolerance or anergy (Mondino and Jenkins 1994; June et al. 1990). Two co-stimulatory molecules in particular, B7.1 (CD80) and B7.2 (CD86), the ligands for CD28 and CTLA-4, have recently received a great deal of attention as potent co-stimulators for T cell function. In humans, B7 is expressed on professional APCs including dendritic cells and is induced on activated B cells, T cells, NK cells and macrophages (Azuma et al. 1993; Freeman et al. 1989). Northern analysis for mRNA expression of B7 revealed that most carcinomas, leukemias of B cell origin (including non-T cell acute lymphocytic leukemia, ALL), prolymphocytic leukemia, hairy cell leukemia and chronic lymphocytic leukemia were B7 negative while some non-Hodgkin's lymphomas were positive (Freeman et al. 1989). These results suggest that lack of B7 expression by many tumors may contribute to their poor immunogenicity. We have confirmed that the majority of colon cancer cells, including those selected for use in the allogeneic vaccine proposed for this protocol, are B7 negative.

In previous studies, transfection of the B7.1 gene into murine melanoma and sarcoma models caused the transfected tumors to be rejected in vivo (Townsend and Allison 1993; Sivasubramanian et al. 1993). In both cases, once immunity was induced, the animals were protected from challenge with the unmodified tumor. Since this immunity was dependent on the presence of cytolytic T cells, it appears that the presence of B7 on the tumor is critical for T cell induction but not for effector cell function. These studies also suggest that the absence of appropriate co-stimulatory molecules on tumors could be a critical factor allowing escape from immune attack despite the expression of potentially strong tumor associated antigens. In more recent studies, combined cytokine (including IL-2) and B7.1 gene transfer have demonstrated synergistic effects in generating efficacious anti-tumor immunity in animal tumor models (Hollingsworth et al. 1995).

11.8 Allogeneic Tumor Cell Vaccines

Immuno-gene therapy would be more practical if allogeneic cells could be employed for immunizations, obviating the need to establish primary fibroblast and colon tumor cultures for each patient. In a number of clinical protocols, allogeneic tumor cells have been used as primary components of immunotherapeutic treatments for brain, skin and breast cancers. These vaccination regimens have been shown to be safe and to generate humoral and cellular anti-vaccine immune responses. Bigner and co-workers have used viable, HLA-mismatched glioma cell lines to vaccinate patients with malignant gliomas (Bullard et al. 1985). They reported the treatment to be safe. In none of the surgical or autopsy examinations was there any evidence of an autoimmune reaction in the brain (encephalomyelitis). Humoral immune responses against the cancer cell line vaccine were observed in a subset of patients. More recently, Belli et al. (1997) immunized metastatic melanoma patients with irradiated, HLA-matched, IL-2-secreting, allogeneic melanoma tumor cell lines. They reported no significant local or systemic side-effects of vaccination. Finally, Smith and co-workers (1997) have immunized breast cancer patients using a CD80/B7.1 genetically modified, HLA-matched, HER2/neu+ allogeneic breast cancer vaccine and granulocyte/macrophage colony-stimulating factor (GM-CSF). The toxicities associated with vaccination, flu-like symptoms and bone pain, were mostly mild (grade 1 or 2) and probably GM-CSF related. Pretreatment skin tests of HLA-matched patients showed that three out of seven responded to unmodified tumor cells whereas six of seven patients responded to B7.1-modified tumor cells.

The rationale for the use of allogeneic tumor cells is predicated upon the expression of shared tumor associated antigens (TAAs) between the tumor used for immunization and the patients' tumors. Several studies have indicated that HLA-A1, HLA-A2 and HLA-A3 haplotypes play a major role in shared TAA presentation which can mediate MHC-restricted tumor destruction by cytolytic T cells (CTLs) (Crowley et al. 1990, 1991; Pandolfini et al. 1991; Chen et al. 1994). In addition, the HLA-A1, HLA-A2 and HLA-A3 haplotypes are relatively common, being expressed by approximately 25%, 50% and 20% of the North American population, respectively. Several shared tumor TAAs defined by CTLs have been described in colon carcinomas (Finn 1993; De Plaen et al. 1994). The protein components of tumor mucin (MUC-1) and the MAGE gene family are TAAs expressed by many colon carcinomas and other adenocarcinomas (Finn 1993; De Plaen et al. 1994). Additional TAAs expressed by the majority of colon carcinomas include carcinoembryonic antigen (CEA) and the glycoprotein recognized by the monoclonal antibodies CO-17-1A and GA733 (Herlyn et al. 1991, 1994).

11.9 Future Clinical Studies

Overall, these pre-clinical and clinical findings provide support for evaluating immuno-gene therapy in colon cancer patients with a mixture of allogeneic tumor cells genetically modified to express B7.1 (CD80) and IL-2-secreting fibroblasts. We have recently submitted for regulatory approvals a Phase I clinical trial in recurrent colon carcinoma patients evaluating the safety of multiple intradermal immunizations with irradiated allogeneic HLA-A2 tumor cell lines (SW620/B7.1, Colo 205/B7.1 and SW403) mixed with irradiated allogeneic fibroblasts (KMST-6/IL-2) genetically modified to express the gene for IL-2. This is an open label, Phase I, single center, multiple dose, dose escalating trial. The allogeneic tumor cell dose will be constant at 6×10^7 irradiated tumor cells (2×10^7 cells per line). The fibroblasts genetically modified to express IL-2 will be dose escalated to provide 0, 400 and 4000 BRMP units of IL-2 per 24 h. The patients will receive three intradermal immunizations at weeks 0, 2 and 4. Twelve patients will be enrolled; four in each dose group. The results of this study should provide useful insights into the efficacy of allogeneic vs autologous tumor cell preparations and the utility of B7.1 (CD80) and IL-2 gene therapy in the treatment of cancer.

References

Azuma M, Yssel H, Phillips JH et al. (1993) Functional expression of B7/BB1 on activated T lymphocytes. J Exp Med 177:845–850

Belli F, Arienti F, Sule-Suso J, Clemente C, Mascheroni L, Cattelan A, Sanatonio C, Gallino GF, Melani C, Rao S, Colombo MP, Maio M, Cascinelli N,

Parmiani G (1997) Active immunization of metastatic melanoma patients with interleukin-2-transduced allogeneic melanoma cells: evaluation of efficacy and tolerability. Cancer Immunol Immunother 44(4):197–203

Borden EC, Sondel PM (1990) Lymphokines and cytokines as cancer treatment. Immunotherapy realized. Cancer 65 [3 Suppl]:800–814

Bubenik J, Viotenok NN, Kieler J, Prassolov VS, Chumakov PM, Bubenikova D, Simova J, Jandlova T (1988) Local administration of cells containing an inserted IL-2 gene and producing IL-2 inhibits growth of human tumors in nu/nu mice. Immunol Lett 19:279–282

Bullard DE, Thomas DGT, Darling JL, Wikstrand CJ, Diengdoh JV, Barnard RO, Bodmer JG, Bigner DD (1985) A preliminary study utilizing viable HLA mismatched cultured glioma cells as adjuvant therapy for patients with malignant gliomas. Br J Cancer 51:283–289

Chen Q, Smith M, Nguyen T, Maher DW, Hersey P (1994) T cell recognition of melanoma antigens in association with HLA-A1 on allogeneic melanoma cells. Cancer Immunol Immunother 38(6):385–393

Crowley NJ, Slinghuff CL, Darrow T et al. (1990) Generation of human autologous tumor specific cytotoxic T cells using HLA-A1 matched allogeneic melanoma. Cancer Res 50:492

Crowley NJ, Darrow TL, Quinn-Allen MA et al. (1991) MHC-restricted recognition of autologous melanoma by tumor-specific cytotoxic T cells. Evidence for restriction by a dominant HLA-A allele. J Immunol 146:1692–1699

De Plaen E, Arden K, Traversari C et al.. (1994) Structure, chromosomal localization and expression of 12 genes of the MAGE family. Immunogenetics 40:360–369

Fakhrai H, Shawler DL, Gjerset R et al. (1995) Cytokine gene therapy with interleukin-transduced fibroblasts: effects of IL-2 dose on anti-tumor immunity. Hum Gene Ther 6:591–601

Fearon ER, Pardoll DM, Itaya T, Golumbek P, Levitsky HI, Simons JW, Karasuyama H, Vogelstein B, Frost P (1990) Interleukin-2 production by tumor cells bypasses T helper function in the generation of an anti-tumor reponse. Cell 60:387–403

Finn OJ (1993) Tumor-rejection antigens recognized by T lymphocytes. Curr Opin Immunol 5:701–708

Freeman GJ, Freedman AS, Segil JM et al. (1989) B7, a new member of the Ig superfamily with unique expression on activated and neoplastic B cells. J Immunol 143:2714–2722

Gandolfi L, Solmi L, Pizza GC, Bertoni F, Muratori R, DeVinci C, Bacchini P, Morelli MC, Corrado G (1989) Intratumoral echo-guided injection of interleukin-2 and cytokine-activated killer cells in hepatocellular carcinoma. Hepatogastroenterology 36:352–356

Gansbacher B, Zier K, Daniels B, Cronin K, Bannerji R, Gilboa E (1990) Interleukin-2 gene transfer into tumor cells abrogates tumorigenicity and induces protective immunity. J Exp Med 172:1217–1223

Herlyn D, Linnenbach A, Koprowski H, Herlyn M (1991) Epitope- and antigen-specific cancer vaccines. Int Rev Immunol 7(4):245–257

Herlyn D, Harris D, Zaloudik J, Sperlagh M, Maruyama H, Jacob L, Kieny MP, Scheck S, Somasundaram R, Hart E et al. (1994) Immunomodulatory activity of monoclonal anti-idiotypic antibody to anti-colorectal carcinoma antibody CO17-1A in animals and patients. J Immunother 15(4):303–311

Hollingsworth S, Gaken J, Darling D et al. (1995) Induction of tumor rejection by combination B7.1/IL-2 expressing tumor cells. Cancer Gene Therapy 2:240

Hoover HC, Brandhorst JS, Peters LC et al. (1993) Adjuvant active specific immunotherapy for human colorectal cancer: 6.5-year median follow-up of a phase III prospectively randomized trial. J Clin Oncol 11:390–399

June CH, Ledbetter JA, Linsley PS et al. (1990) Role of the CD28 receptor in T cell activation. Immunol Today 11(6):211–216

Kelso A (1989) Cytokines: structure function and synthesis. Curr Opin Immunol 2(2):215–225

Kim TS, Cohen EP (1994) Interleukin-2-secreting mouse fibroblasts transfected with genomic DNA from murine melanoma cells prolong the survival of mice with melanoma. Cancer Res 54(10):2531–2535

Kim TS, Russell SJ, Collins MK, Cohen EP (1992) Immunity to B16 melanoma in mice immunized with IL-2-secreting allogeneic mouse fibroblasts expressing melanoma-associated antigens. Int J Cancer 51(2):283–289

Lotze MT, Chang AE, Seipp CA et al. (1986) High-dose recombinant interleukin-2 in the treatment of patients with disseminated cancer: responses, treatment-related morbidity and histologic findings. JAMA 256:3117–3124

Mondino A, Jenkins MK (1994) Surface proteins involved in T cell costimulation. J Leukoc Biol 55(6):805–815

Pandolfini F, Boyle LA, Tretin L et al. (1991) Expression of HLA-A1 antigen in human melanoma cell lines and its role in T cell recognition. Cancer Res 51:3164–3170

Pizza G, Viza D, DeVince C, Vichi-Pascuuchi JM, Busutti L, Bergami T (1988) Intralymphatic administration of interleukin-2 (IL-2) in cancer patients: a pilot study. Cytokine Res 7:45–48

Rosenberg SA, Lotze MT, Mule JJ (1988) New approaches to the immunotherapy of cancer. Ann Intern Med 108:853–864

Sarna G, Collins J, Figlin R, Robertson P, Altrock B, Abels R (1990) A pilot study of intralymphatic interleukin-2. II. Clinical and biological effects. J Biol Response Modif 9:81–86

Shawler DL, Dorigo O, Van Beveren C, Bartholomew RM, Fakhrai H, Sobol RE (1997) Comparison of interleukin-2 (IL-2) gene therapy with allogeneic fibroblasts in the CT-26 model of murine colorectal carcinoma. Oncol Rep 4:135–138

Sivasubramanian B, Ostrand-Rosenberg S, Nabavi N et al. (1993) Constitutive expression of B7 restore immunogenicity of tumor cells expressing truncated major histocompatibility complex class II molecules. Proc Natl Acad Sci USA 90:5687–56890

Smith JW II, Schoof DD, Disis ML, Brant-Zawadski P, Wood W, Doran T, Johnson E, Urba WJ (1997) Genetic immunization of patients with metastatic breast cancer using a CD80 (B7.1)-modified, HLA-A2+, HER2/neu+, allogeneic breast cancer cell vaccine plus GM-CSF. Cancer Gene Ther 4(6):S48

Sobol RE, Fakhrai H, Shawler DL et al. (1995) Interleukin-2 gene therapy in a patient with glioblastoma. Gene Ther 2:164–167

Tahara H, Zeh HJ III, Storkus WJ et al. (1994) Fibroblasts genetically engineered to secrete interleukin 12 can suppress tumor growth and induce antitumor immunity to a murine melanoma in vivo. Cancer Res 54(1):182–189

Tepper RI, Pattengale PK, Leder P (1989) Murine interleukin-4 displays potent anti-tumor activity in vivo. Cell 57:503–512

Townsend SE, Allison JP (1993) Tumor rejection after direct costimulation of $CD8^+$ T cells by B7-transfected melanoma cells. Science 259:368–370

Veelken H, Mackensen A, Lahn M; Kohler G, Becker D, Franke B, Brennscheidt U, Kulmburg P, Rosenthal FM, Keller H, Hasse J, Schultze-Seemann W, Farthmann EH, Mertelsmann R, Lindemann A (1997) A phase-I clinical study of autologous tumor cells plus interleukin-2-gene-transfected allogeneic fibroblasts as a vaccine in patients with cancer. Int J Cancer 70(3):269–277

Watanabe Y, Kuribayashi K, Miyatake S, Nishihara K, Nakayama EL, Taniyama T, Sakata TA (1989) Exogenous expression of mouse interferon gamma cDNA in mouse neuroblastoma C1300 cells results in reduced tumorigenicity by augmented anti-tumor immunity. Proc Natl Acad Sci USA 86:9456–9460

12 Gene Therapy Regulatory Issues in the United States and Europe

R. Wikberg-Leonardi

12.1 Introduction

In this chapter I will give a brief overview of initial interactions with regulatory agencies during the development of a new therapy, for example a gene therapy product. The derivation of the rules in the United States and how gene therapy products fit into this will be described. In addition I will discuss interactions with the Food and Drug Administration (FDA), highlighting certain points to keep in mind. Finally a brief overview of European regulatory interactions for gene therapy products will also be provided.

12.2 United States

12.2.1 Historic Background on Regulations Relative to Drugs and Biologics

Historically, drugs and biologics have been regulated separately. Today there is a significant move to narrow the differences between how a drug is evaluated compared to a biologic. What is considered a drug? And what is a biologic? The first part of the question seems easy – most chemical, synthesized products are categorized as drugs. Biologics are vaccines, blood and blood products, serums, toxins, antitoxins, organic arsenic compounds and allergenic compounds. Some may think that natural products or recombinant products always fall into the biologics category, but this is not necessarily true. There are many products for which the category of drug or biologic is not so clear cut.

Most laws and regulations relative to drugs and biologics have evolved out of major health crises, health tragedies, or from fraud. While drug regulations evolved from the US Department of Agriculture (USDA) and its laws, biologic regulations evolved from public health concerns and its laws.

12.2.2 Drugs

Until 1906, there was no comprehensive federal statute regulating drug products in the US. In 1906 Congress passed the Pure Food and Drugs Act, a weak law, which prohibited the interstate transportation of adulterated and misbranded foods and drugs. This act was initially administered by the USDA. The FDA, partially as we know it today, was established in 1931.

The Federal Food Drug and Cosmetic (FD&C) Act, much stronger than the 1906 act, was established in 1938. The impetus to enactment can be attributed to the 1937 "Elixir Sulfanilamide" tragedy. The product was not truly an elixir, but actually used diethylene glycol as a solvent; the safety of the product was not tested prior to its marketing. The 1938 FD&C Act required pre-market safety testing for approval. However, it was not until 1962, following the thalidomide tragedy, that pre-market approval for efficacy of drugs was introduced.

12.2.3 Biologics

The first and very strong biologics law was The Virus, Serum and Toxin Act, which was enacted in 1902 following a tragedy with a diphtheria vaccine that was contaminated with tetanus. This law provided a regulatory scheme for viruses, sera, toxins, antisera and analogous products as well as a licensing process giving inspection authority (principles of safety, purity and potency). This licensing process continues today in the Establishment License and Product License that must be obtained for the manufacture of a licensed biological product.

Today biologics are regulated under the Public Health Service Act, established in 1912 and amended and recodified in 1944 to add the 1902 Virus, Serum and Toxin Act. The Marine Fund was established in 1798 to fund research in biologics and ultimately had the responsibility to administer any existing regulation and the 1902 act. This was subsequently taken over by the Public Health Service when biologics review was part of the National Institute of Health (NIH) (Division of Biological Standards), from 1930 through 1972, after which the FDA Bureau of Biologics was created to administer the regulation.

Gene therapy products are defined as products containing genetic material administered to modify or manipulate the expression of genetic material or to alter the biological properties of living cells. Gene therapy products that are not devices, especially those containing viral vectors to be administered to humans, fall within the definition of biological products and are subject to the licensing provisions of the Public Health Act, as well as to the drug provisions of the FD&C Act such as conformity with current Good Manufacturing Practices (cGMP) regulations (Federal Register, vol. 58, 197, 10/14/93, 53249).

Early gene therapy development focused on treatment of life threatening and severely debilitating illnesses. Drugs aimed at, for example, cancer treatment are often eligible for an expedited review by the FDA

(21CFR312 subpart E). Initiation of early discussions with the FDA for these kinds of products is important.

In the early development of gene therapy the NIH played a significant role through the Recombinant DNA Advisory Committee (RAC). Review of a clinical gene therapy protocol was obligatory if it had NIH funding, and, although the RAC did not have any statutory authority to approve or disapprove clinical protocols, considerable time was spent by sponsors in discussions with RAC to obtain "approval" and with the FDA to ultimately get approval for initiating the clinical program. Today the RAC mandate and charter have been reorganized; the RAC will continue to review and discuss novel human gene transfer experiments but without authority to approve or disapprove clinical trials. This will still allow public access to human gene transfer clinical trial information (Federal Register, vol. 61, 227, 11/22/96, 59725).

12.2.4 The Pre-IND Meeting

Following this short summary of the historic development of the laws, I will discuss initial gene therapy submissions to the FDA and interactions with the Center for Biologics Evaluation and Research (CBER) during the development phase of the gene therapy product. At what stage of the research and development (R&D) program should a pre-IND (Investigational New Drug Applications) meeting be considered and why is a meeting with the FDA important?

The plans for a pre-IND discussion with the FDA should begin when preclinical pharmacology studies show proof of principle. The pre-IND meeting request to FDA must contain background information on the pre-clinical development performed and planned. It is important to outline the development plan for Chemistry, Manufacturing and Controls, including tentative specifications for the clinical product as well as plans for the safety/toxicology studies planned. In addition an outline of the clinical plan for the initial clinical study should be included.

The pre-IND meeting is an opportunity for the sponsor to get the FDA's intellectual by-in to (i.e., acceptance of) the program. If the product being developed is to be classified or be categorized for expedited review, this is also the perfect opportunity to initiate the discussions with the FDA.

It is extremely important that the team going to the meeting is very well rehearsed. Meetings with CBER are very productive. In many areas there is an open-door policy with a, "come and talk to us about your project" attitude. However, the people at CBER have been burnt by sponsors who are not prepared. Meetings have been requested, but background information has not been provided in a timely manner. When the information finally is received, scheduled meetings have been canceled due to insufficient information contained in the meeting background package.

Following the meeting, it is important to exchange minutes of the meeting with the agency. This avoids future misinterpretations of the agreements reached during the meeting.

12.2.5 IND Preparation

Each section of an IND for a Phase 1 clinical study is presented below, followed by a comment on contents, and details of data and length based upon the November 1995 guideline (Guidance for Industry, Content and Format of Investigational New Drug Applications (INDs) for Phase 1 Studies of Drugs, Including Well-Characterized Therapeutic, Biotechnology-Derived Products, November 1995).

- *Items 1 and 2: Table of Contents and Form FDA 1571.* The form is used for the initial submission, as well as for every subsequent IND amendment.
- *Items 3 and 4: Introductory Statement and General Investigational Plan.* These two sections should be kept brief, at most two to three pages. The FDA is only looking for plans for the first year of the IND. Detailed development plans cannot be properly written before the results of the Phase 1 study are known.
- *Items 5 and 6: Investigator's Brochure and Phase I protocol.* The development of the Investigator's Brochure, which provides an overview of the relevant information and scientific results known about the drug, should be written following the International Committee on Harmonization (ICH) guideline (ICH Good Clinical Practice: Consolidated Guideline).

A Phase 1 study is commonly a safety study performed in healthy human subjects. For most gene therapy products, the first clinical study is one in patients and could be called a Phase 1–2 study. The regulations state that Phase 1 protocols should be directed primarily at providing an outline of the clinical investigation; an estimated number of subjects to be studied; a description of safety exclusion criteria; a description of the dosing plan including duration, dose, or method to be used in determining dose. Only criteria that are critical to the subjects safety need to be specified, i.e., monitoring of vital signs and blood chemistries, toxicity-based stopping or dose adjustment rules. Please note that modifications to a Phase 1 protocol are required to be reported to the FDA only in the IND annual report. Please note that CBER requires submission of investigator information (FDA Form 1572 and CV), Institutional Review Board (IRB) approval and IRB-approved Informed Consent prior to initiation of the clinical study.

- *Item 7: Chemistry, Manufacturing and Controls.* The guideline for Phase 1 studies discusses the Chemistry, Manufacturing and Controls (CMC) information to be submitted at this early stage in the development of the product. In each phase of the clinical development stage, sufficient information should be submitted to assure proper identification, quality, purity and strength of the drug product. It is recognized that modifications to the manufacturing of the product may occur during the development. For the initial submission, the most important aspect is that sufficient information is included to identify any safety concerns for the product to be used in the clinic. There are several Points to Consider and Guidance documents relative to gene therapy products that should be used as well as ICH guidance documents. According to information from CBER, a specific Guidance Document for Chemistry, Manufacturing and Controls Information for Gene Therapy Products is being developed. No draft was available at this time. Please note that CBER requires clinical batch release data to be submitted to the IND prior to use in the clinic.
  - *Guidelines for Chemistry, Manufacturing and Controls:*
    Draft Points to Consider in the Characterization of Cell Lines Used to Produce Biologicals
    Guidance in Human Somatic Cell Therapy and Gene Therapy

ICH Derivation and Characterization of Cell Substrates used for Production of Biotechnological/Biological Products
ICH Guidelines on Stability Testing of Biotechnological/Biological Products

- *Item 8: Pharmacology and Toxicology Data.* This section contains the Pharmacology and Toxicology Summary, In Vitro Studies, Pharmacology Study Reports and Toxicology Study Reports. The integrated summary of pharmacology and toxicology studies and results are an important part of the IND. Results from pharmacology, safety and toxicology findings must be discussed here. Since no standard safety/toxicology testing program for gene therapy products exists, the pre-IND discussions with the FDA are important. In early recombinant protein development and gene therapy development, animal studies that really were irrelevant were performed. If a good pre-clinical pharmacology model exists, it probably is the best animal model to be used for the safety testing. Additionally, the FDA prefers to gather as much safety data as possible in the pharmacology studies. The evaluation of preclinical safety is the basis for the safety in the proposed clinical study. It is proposed that the summary, including both pharmacology and toxicology, be about five to ten pages.

It is possible to make the IND submission with draft Good Laboratory Practice (GLP) toxicology study reports. If unaudited reports are used for the IND submission a safety update must be submitted to the IND within 120 days, discussing any differences in findings between the unaudited reports and the final reports. If no differences are found this should be stated.

- *Item 9: Previous Human Experience.* Here, any previous human experience relevant to the new gene therapy product to be tested should be stated. If not, this should also be stated.

Initiation of the clinical study must wait for CBER review. When the IND is filed, the Form FDA 1571 that is part of the IND filing contains the following text: "I agree not to begin clinical investigations until 30 days after FDA's receipt of the IND unless I receive earlier notification by FDA that the studies may begin. I also agree not to begin or continue clinical investigations covered by the IND if those studies are placed on clinical hold. I agree that an Institutional Review Board (IRB) that complies with the requirements set forth in 21CFR, Part 56 will be responsible for the initial and continuing review and approval of each of the studies in the proposed clinical investigation. I agree to conduct the investigation in accordance with all other applicable regulatory requirements."

12.2.6 Interactions with the Center for Biologics Evaluation and Research

Continued interaction with the FDA is important during the development phase of the gene therapy product. For a new gene therapy with safety as the major focus in the initial clinical phase, frequent interactions during the clinical development are mandatory. In addition, there are several so-called codified meetings, the End of Phase 2 meeting and pre-Biologics License Application (BLA) and Establishment License Application (ELA) submission meetings. Continued exchange and shared information between the sponsor and FDA will benefit the approval process, but most important is that the science performed is sound and good.

In efforts by the US government to modernize the review process, the terminology, "well-characterized biological" has been defined. The well characterized biologicals consist of monoclonal antibodies and recombinant proteins. For this category of biologics no separate ELA needs to be filed and the BLA application resembles a New Drug Application (NDA).

The FDA Modernization Act has added a new section to the FD&C Act to expedite the development and approval of new drugs that address unmet medical needs relating to serious or life-threatening conditions. These drugs can be approved based upon surrogate end-points and clinical end-points. Of course the discussions with the FDA at the pre-IND stage are very important in reaching agreement regarding these end-points. One of the provisions of this revised law is a "rolling" NDA/BLA review. The FDA may begin reviewing portions of the NDA/BLA as the documentation is ready, rather than waiting for the complete application to be submitted and filed. A guidance document regarding the fast track procedure is currently being developed by the FDA.

12.3 Europe

12.3.1 Regulatory Environment

A new European centralized system for the authorization of medicinal products was implemented in 1995. The European Agency for the Evaluation of Medicinal Products (European Medicines Evaluation Agency, EMEA) is located in London. The marketing application for a gene therapy product must be submitted via the centralized procedure. Individual country-by-country applications cannot be made for a new marketing application for a biotechnology product.

However, at the clinical study phase each European country has its own regulations. Most countries have a drug review and a gene therapy review that must be done before initiation of clinical studies. For example, in the UK a CTX (clinical trial exemption) must be submitted to the MCA (Medicines Control Agency); in addition, the clinical protocol must be submitted to the Gene Therapy Advisory Committee (GTAC). The review time of a CTX is approximately 45 days, but can often be up to 60 days. The GTAC schedules official meetings four to six times per year and requests that protocols be submitted 90 days prior to a scheduled meeting. Therefore, at this time it seems easier to do the initial clinical gene therapy study in the United States and approach the EMEA in the early development stage, when some initial clinical data are available.

12.3.2 Organization of European Medicines Evaluation Agency

The EMEA is in charge of coordinating regulatory scientific resources existing in the 15 member states with a view to evaluating and supervising medicinal products for both human and veterinary use. The EMEA consists of a management board with two representatives from each of the 15 member states, two scientific committees, CPMP (Committee for Proprietary Medicinal Products) and CVMP (Committee for Veterinary Medicinal Products), and the secretariat located in London.

One of the EMEA priorities, determined by the Management Board and listed in the 1997 work book, is scientific advice to future applicants and the European Union (EU) institutions. Since the inception of the EMEA, the number of meetings with sponsors has steadily increased.

The EMEA is willing to meet with companies to discuss their projects, very much in the same way as an FDA pre-IND meeting is held. The specific scientific input for a gene therapy product and the development plan will come from the CPMP Biotechnology Working Party (BWP). The BWP meet eight or nine times a year, in order to allow it to support the preparation of opinions for the CPMP regarding biotechnology products. They also, especially for gene therapy products, advise that a summary package with information be submitted early during the development to the CPMP BWP with specific questions from the sponsor regarding the development. Since I proposed that initial human Phase 1–2 data be obtained in the US, the appropriate time to approach the EMEA is at the end of the Phase 1–2 study.

The group responsible for Evaluation of Human Medicines within EMEA, and particularly the group responsible for biotechnology and biologicals, is willing to help the sponsor with the initial review of the background summary information. They will work with the company in the preparation of the specific questions to be put to the CPMP. This is an excellent opportunity to get consolidated European comments on the adequacy of development plans for toxicology studies, proposed specifications and clinical development plans.

12.3.3 Filing for Marketing Approval

Prior to filing a Marketing Application for a gene therapy product in Europe, a Rapporteur and a Co-rapporteur will be assigned to the filing. The role of the Rapporteur is basically to be the lead reviewing country of the Marketing Application. The sponsor will have an opportunity to request the Rapporteur country of choice, but the request is not necessarily granted. Most sponsors will request one of the nations with Health Authorities with scientific reviewers able to give scientifically good review of an application and also meeting established timelines in the review process, like the UK, The Netherlands and Sweden; thus, the review, , the spread of reviews and work load is not sufficient. The Rapporteur and Co-rapporteur are assigned by the Chairman of the CPMP, currently Professor Alexandre.

The format of the different parts of the Marketing Application is specified in the Rules Governing the Medicinal Product in the European

Community. The application must be partially translated into 15 different languages prior to submission, since every country will get their own copy, not only the Rapporteur and Co-rapporteur countries.

12.4 Conclusion

What is the road to success in the regulatory development of a gene therapy product?? My advice is:

1. Be interactive!
2. Keep the health authorities involved and informed!
3. Establish good contacts!
4. Remember: Good science works!

Subject Index

Ernst Schering Research Foundation Workshop

Editors: Günter Stock
Ursula-F. Habenicht

Vol. 1 *(1991)*: Bioscience ⇆ Society – Workshop Report
Editors: D. J. Roy, B. E. Wynne, R. W. Old

Vol. 2 (1991): Round Table Discussion on Bioscience ⇆ Society
Editor: J. J. Cherfas

Vol. 3 (1991): Excitatory Amino Acids and Second Messenger Systems
Editors: V. I. Teichberg, L. Turski

Vol. 4 (1992): Spermatogenesis – Fertilization – Contraception
Editors: E. Nieschlag, U.-F. Habenicht

Vol. 5 (1992): Sex Steroids and the Cardiovascular System
Editors: P. Ramwell, G. Rubanyi, E. Schillinger

Vol. 6 (1993): Transgenic Animals as Model Systems for Human Diseases
Editors: E. F. Wagner, F. Theuring

Vol. 7 (1993): Basic Mechanisms Controlling Term and Preterm Birth
Editors: K. Chwalisz, R. E. Garfield

Vol. 8 (1994): Health Care 2010
Editors: C. Bezold, K. Knabner

Vol. 9 (1994): Sex Steroids and Bone
Editors: R. Ziegler, J. Pfeilschifter, M. Bräutigam

Vol. 10 (1994): Nongenotoxic Carcinogenesis
Editors: A. Cockburn, L. Smith

Vol. 11 (1994): Cell Culture in Pharmaceutical Research
Editors: N. E. Fusenig, H. Graf

Vol. 12 (1994): Interactions Between Adjuvants, Agrochemical and Target Organisms
Editors: P. J. Holloway, R. T. Rees, D. Stock

Vol. 13 (1994): Assessment of the Use of Single Cytochrome P450 Enzymes in Drug Research
Editors: M. R. Waterman,
M. Hildebrand

Vol. 14 (1995): Apoptosis in Hormone-Dependent Cancers
Editors: M. Tenniswood, H. Michna

Vol. 15 (1995): Computer Aided Drug Design in Industrial Research
Editors: E. C. Herrmann, R. Franke

Vol. 16 (1995): Organ-Selective Actions of Steroid Hormones
Editors: D. T. Baird, G. Schütz, R. Krattenmacher

Vol. 17 (1996): Alzheimer's Disease
Editors: J.D. Turner, K. Beyreuther, F. Theuring

Vol. 18 (1997): The Endometrium as a Target for Contraception
Editors: H.M. Beier, M.J.K. Harper, K. Chwalisz

Vol. 19 (1997): EGF Receptor in Tumor Growth and Progression
Editors: R. B. Lichtner, R. N. Harkins

Vol. 20 (1997): Cellular Therapy
Editors: H. Wekerle, H. Graf, J.D. Turner

Vol. 21 (1997): Nitric Oxide, Cytochromes P 450, and Sexual Steroid Hormones
Editors: J.R. Lancaster, J.F. Parkinson

Vol. 22 (1997): Impact of Molecular Biology and New Technical Developments in Diagnostic Imaging
Editors: W. Semmler, M. Schwaiger

Vol. 23 (1998): Excitatory Amino Acids
Editors: P.H. Seeburg, I. Bresink, L. Turski

Vol. 24 (1998): Molecular Basis of Sex Hormone Receptor Function
Editors: H. Gronemeyer, U. Fuhrmann, K. Parczyk

Vol. 25 (1998): Novel Approaches to Treatment of Osteoporosis
Editors: R.G.G. Russell, T.M. Skerry, U. Kollenkirchen

Vol. 26 (1998): Recent Trends in Molecular Recognition
Editors: F. Diederich, U. Eder

Vol. 27 (1998): Gene Therapy
Editors: R.E. Sobol, K.J. Scanlon, E. Nestaas

Supplement 1 (1994): Molecular and Cellular Endocrinology of the Testis
Editors: G. Verhoeven, U.-F. Habenicht

Supplement 2 (1997): Signal Transduction in Testicular Cells
Editors: V. Hansson, F. O. Levy, K. Taskén

Supplement 3 (1998): Testicular Function: From Gene Expression to Genetic Manipulation
Editors: M. Stefanini et al.